Essentials of Control

Essentials of Control

JOHN SCHWARZENBACH

Senior Lecturer in the Department of Mechanical Engineering
University of Leeds

Longman

Addison Wesley Longman Limited
Edinburgh Gate
Harlow
Essex
CM20 2JE, England
and Associated Companies throughout the world.

First published 1996

British Library Cataloguing in Publication Data
A catalogue entry for this title is available from the British Library.

ISBN 0-582-27347-1

Library of Congress Cataloging-in-Publication Data
A catalog entry for this title is available from the Library of Congress.

Produced through Longman Malaysia, TCP

CONTENTS

PREFACE

Control is a subject which forms part of most programmes of study for degrees and diplomas in mechanical engineering, electrical engineering, chemical engineering and related disciplines. Many students find it daunting and difficult, largely because of its mathematical nature with apparently abstract concepts which cannot be readily visualised. Too often students attempt to remember techniques for solving problems without understanding the significance of what they are doing. It is a subject where the many different topics are all interrelated, and hence where the significance of what is studied in the early stages becomes more fully appreciated later. Often, towards the end of a course, things fall into place and students find that the subject is not really as difficult as they thought.

Although there is a wide choice of available textbooks, students are faced with the facts that most are comparatively expensive and that it is often difficult to get a sound appreciation of the fundamentals, because of comprehensive coverage in which it is not easy to extract the key information from the detail. The objective of this book is to present in a clearly understandable form the basic fundamentals and concepts of control, the essence of the subject, in a slim, affordable text. The book includes brief introductions to some of the more advanced concepts and techniques. For the student following a single module or course based primarily on classical linear theory the book should serve well as a supportive course text. If a module is based predominantly on modern control theory then the book should help the student gain understanding of system behaviour and control. Where control needs to be studied in greater detail or to a more advanced level then the understanding gained through this text should serve as a good basis for reference to more detailed or advanced textbooks.

PRINCIPAL SYMBOLS

The symbols listed here appear relatively widely in the book. The significance of others which appear only locally should be clear from the accompanying text and figures.

t	the independent variable time
$x(t)$, $y(t)$, $v(t)$, $i(t)$, $\theta(t)$, etc	variables which are functions of time (normally written lower case)
s	the Laplace operator; $s = \sigma + j\omega$, a complex quantity (where σ is the real part and ω the imaginary part)
$\mathscr{L}$	the Laplace transform of ...
$X(s)$, $Y(s)$, $V(s)$, $I(s)$, $\theta(s)$, etc	variables which have been transformed into the s domain (normally written upper case)
subscript i, o	input and output variables
$G(s)$	a transfer function, normally in a forward path
$G_p(s)$, $G_c(s)$	process and controller transfer functions
$H(s)$	a transfer function in a feedback path
$R(s)$, $E(s)$, $C(s)$	reference input, error, and controlled output signals
m, k, c,	mass, spring stiffness, and viscous damping coefficient
J	inertia of rotational load
R, L, C	resistance, inductance, capacitance
K	steady state gain
τ	time constant (of first order component)
ζ, ω_n	damping factor, undamped natural frequency (of second order component, or associated with a pair of complex conjugate roots)
p_1, p_2, ...	poles of transfer function, roots of characteristic equation
z_1, z_2, ...	zeros of transfer function
$G(j\omega)$	sinusoidal transfer function
$\lvert G(j\omega) \rvert$, $\underline{/G(j\omega)}$	magnitude and phase of $G(j\omega)$, the ratio of output amplitude to input amplitude, and the phase shift between input and output sine waves

db	units of decibels (20 times $\log_{10}$ of quantity)
GM, PM	gain margin and phase margin
M, ϕ	closed loop magnification and phase
M_p, ω_p	peak closed loop magnification and frequency at which it occurs
K_p, K_v, K_a	positional, velocity and acceleration constants
P+I+D	proportional + integral + derivative control
ϕ_m, ω_m	maximum phase lead from phase lead network and frequency at which it occurs
α	ratio of time constants in phase lead ($\alpha > 1$) and phase lag ($\alpha < 1$) networks

CHAPTER 1

CLOSED LOOP SYSTEMS AND THEIR MODELS

This chapter starts by describing the general nature of control, and explains the principles of open loop control and of closed loop or feedback control, with simple examples from everyday life. Analysis and design of a control system requires a mathematical model, this being basically a set of differential equations describing the relationships between the system variables, usually supplemented by pictorial representation in the form of a block diagram. Section 1.2 is a descriptive introduction to mathematical models, block diagrams and the concept of linearity. In classical control systems analysis the manipulation and solution of the differential equations is effected by use of the Laplace transform technique, with the differential equations written in the form of transfer functions. Section 1.3 outlines the Laplace transform technique in sufficient depth to allow it to be used with understanding as a mathematical tool, and then defines the transfer function, and shows how it is derived from the governing differential equations. In Section 1.4 transfer functions are derived for a number of simple physical systems to illustrate the method of approach, the simplifying assumptions which must be made, and the form of the resulting transfer functions. Finally, Section 1.5 illustrates the ease with which the system equations can be manipulated by the process of block diagram reduction.

1.1 Open and closed loop control

Consider first the nature of the subject of control engineering or control systems, often referred to simply as 'control'. It is concerned broadly with the analysis and design of systems to control particular variables in dynamic processes in some desired way. The two main requirements are that of maintaining a variable sensibly constant and close to a desired value in spite of disturbances to the process, and that of altering a variable to a new desired value quickly and accurately yet without excessive overshoot or oscillation. An example of the former is that of maintaining the inside of an oven or a freezer close to the desired temperature value in spite of variations in ambient temperature, of the door being opened to inspect the contents, or of objects at ambient temperature being placed inside. The control system must adjust the amount of heating or cooling to minimise the deviation from the desired temperature. Examples of the latter are the need for fast and accurate positioning of the read/write head for a computer disc or the laser head for a CD player, and the need for the limbs of a robot to be moved in such a way that the tool or component being manipulated is moved along some desired path with sufficient accuracy.

In the subject of dynamics a mechanical system involving inertia, springs and damping would be studied to determine how it is expected to behave dynamically – how the position of any chosen point on the mechanism varies as a function of time when the system is disturbed by the action of either a time varying position or force. Design is then a process of selecting appropriate parameter values to achieve acceptable dynamic behaviour, and once designed, manufactured and assembled there is no attempt during operation to monitor the behaviour and to make adjustments should some unexpected disturbance cause an unwanted deviation.

Control is concerned with dynamics in a wider sense – with seeking understanding of the dynamic behaviour of mechanical, electrical, chemical, biological, economic and other forms of system with a view to controlling in some desired manner variables such as position, velocity, current, voltage, temperature, pressure, flow rate, chemical concentration, and currency exchange rate. One form of 'control' which is not included is **sequential control**, a process used in automation and in many 'automatic' systems such as washing machines, wrapping machines, cash dispensers, and passenger lifts. Each of these is characterised by there being a sequence of operations with, in general, the conclusion of one operation signalling the start of the next. Sometimes a branching sequence with an element of logic is involved. Factors such as inertia and damping are likely to influence only the duration of the stages of the sequence, and not to any significant extent their nature. At the design stage, modes of possible malfunction should be anticipated and measures taken to cope with eventualities in a sensible way, e.g. a bottling machine should not attempt to fill a missing bottle, an automatic drilling machine should not attempt to drill an incorrectly positioned component, a passenger lift should not move if overloaded. In contrast, control theory and practice is concerned with the control of system variables in a continuous analogue manner, e.g. control of a heat source in order to maintain a temperature at a chosen set value, or to increase the temperature at a defined rate of change, as opposed to simply switching off the heat once a set temperature is reached or after a given time.

There are two approaches to control—open loop control, which because of its shortcomings has limited usefulness, and closed loop control which has wide application and is the subject of this book. A simple dynamic system or process can often be considered to have one input variable, which when altered causes the state of the system to change, and one output which is the variable of interest ('cause' and 'effect'). There may, in addition, be disturbance inputs over which one has no influence. In the absence of disturbance, if a suitable mathematical model (equations describing the relationships between the variables) is available then it is possible to determine what input function would be needed to achieve a certain desired output, and to apply this signal to the process. This is **open loop control**, often referred to as **scheduling control**, Fig. 1.1(a). There is no mechanism to correct for errors in the output arising from unknown disturbance inputs, or for errors in or changes to the assumed model. Once a system has been designed and built then good control is dependent on being able to anticipate disturbances and model changes and determine a suitable input function. With most processes control is potentially improved if the output is continually monitored and compared with the desired output value, and if the difference

Fig. 1.1 (*a*) Open loop control; (*b*) closed loop control

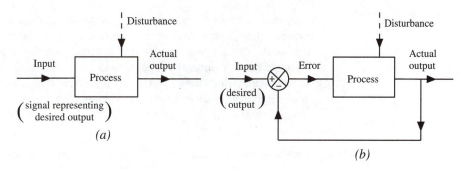

between the two (the error) is used as a signal to develop corrective action to reduce the error. This is the basis of **closed loop control** or **feedback control**, Fig. 1.1(b). It is widely used, and potentially permits significantly better control in that there is a degree of automatic correction for changes arising from disturbances and model changes, but accompanying this is a tendency to oscillatory and perhaps even unstable behaviour. Sometimes with closed loop control the desired output is a constant value, and the aim is to maintain the actual output as close as possible to the desired value – such a system is a **regulator**. Sometimes it is desired that the output changes in some particular time dependent way, in which case the objective is that the actual output should follow as closely as possible some desired time function – the system then being referred to as a **servomechanism**, frequently abbreviated to **servo**. The same feedback loop will satisfy both requirements.

Before becoming involved in mathematical analysis, some general applications of open and closed loop control are outlined.

(a) Simple actions, such as sweetening coffee, diluting fruit juice, baking a cake, draw on past experience when aiming for a desired result. Moderate deviation from the ideal is acceptable; continuous feedback is unnecessary.

(b) An electric toaster requires past experience to decide on the appropriate setting for a given type of bread, and an automatic washing machine to be efficiently used requires the user to judge the appropriate programme for the wash load. To measure the toast quality or fabric cleanliness for feedback control would be relatively difficult, and would involve unwarranted expense.

(c) The domestic oven, central heating system, refrigerator, and cold water cistern are examples where feedback is used to control temperature or liquid level. The output signal is continually monitored and an error signal generated in a simple inexpensive manner; this signal then determines the amount of heat or fluid flow needed.

(d) Many industrial processes require temperature, pressure, liquid level, flow rate, chemical composition, thickness, light intensity, or some other parameter at certain points in the process to be maintained at near constant values. Feedback loops acting as regulators achieve this.

(e) In industry there are also various processes where position or velocity at some point must be altered in a certain way, e.g. movement of a workpiece past a cutter in a numerically controlled milling machine, control of a robot, control of roller speeds in a printing press at start-up and shut-down, tracking of a hostile aircraft by defence weapon systems. These are servomechanism applications of closed loop control.

(f) The financial world presents examples of both open and closed loop control. Actions such as the changing of tax rates, interest rates offered to lenders and charged to borrowers and fees for educational courses are forms of open loop control action where the use of feedback to continually monitor and adjust is impractical. In the share and money markets feedback and resultant price or exchange rate adjustments are continuous.

(g) Some systems which have traditionally been open loop now sometimes incorporate feedback to meet more exacting specifications, e.g. automatic exposure control in a camera, control of fuel supply to an internal combustion engine, and active suspension systems for vehicles.

In a general sense, measures taken to achieve greater accuracy of control will make a system more oscillatory, and can make it unstable. The engineer must therefore understand the nature of the behaviour of feedback systems, so that a system can be designed with an appropriate compromise between the requirements of accuracy and stability.

1.2 Mathematical models

Analysis of any given system, and the process of design to achieve acceptable dynamic behaviour, requires a suitable mathematical model for the system. A **mathematical model** is an equation or set of equations which defines the relationships between the variables that describe the state of the system as a function of time. For a system comprising more than one functional component it is helpful to supplement the equations by some form of schematic diagram, the most common form and the one used in this book being the **block diagram**. Each functional component is considered to have one signal input and one signal output, and the output of one component frequently is the input to another. In a block diagram labelled boxes represent the individual components and directional lines represent the signal flow paths, with addition or subtraction of signals being effected at summing junctions. In Fig. 1.1 the difference between an open loop and a closed loop system is illustrated by means of simple block diagrams. Figure 1.2 shows a typical block diagram, for a positional servomechanism. This

Fig. 1.2 Descriptive block diagram for positional servomechanism

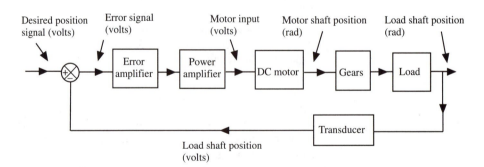

diagram identifies the components which make up the system, how they are connected, and the nature of the input, output, and intermediate signals. In this case the system output is the angular position of a shaft supporting a load, and this is measured by a transducer which outputs a voltage which ideally is exactly proportional to the angle. This voltage is subtracted from an input voltage signal representing the desired output, to create an error voltage. This is amplified by a voltage amplifier (which would also effect the subtraction), and the amplifier output signal further amplified by a power amplifier whose function is to generate a voltage output with sufficient current capability to serve as input to the electric motor which drives the load shaft through some gearing. The larger the difference between the actual and desired output positions, the larger will be the motor torque, and ideally the system will not come to rest until the error is zero. To undertake analysis or design, each signal must be identified by a symbol and each block represented by an equation relating the output to the input.

The reader who has studied dynamics will know that mechanical systems comprising arrangements of components with mass, stiffness, and damping of a viscous nature are described mathematically by differential equations, and that analysis of the motion within the system involves solving these equations. For dynamic systems in a wider sense, the relevant governing equations are also differential equations, as will be illustrated by a series of examples in Section 1.4. If these equations are linear then analysis is significantly more straightforward than if they are non-linear. Many components can be closely represented by linear differential equations over a certain range of operation, and some components which are inherently non-linear can be approximated by linear differential equations provided the range of operations is small. In the early stages of analysis it is generally assumed that a reasonable first estimate of system behaviour can be obtained by using linear equations. Consideration of non-linearity can then be brought in as a refinement at a later stage, and Section 4.4 provides an introduction to the main methods available.

A linear differential equation is one which has constant coefficients and which contains no terms involving products of variables or their derivatives. If a system is linear (described by linear differential equations) then the **principle of superposition** applies, which means that when two forms of input function are applied simultaneously (superposed) then the resultant response is the sum of the responses to the inputs applied individually. A consequence of this is that if the amplitude of a given input or forcing function is altered by some factor k then the form of the response remains unchanged but its magnitude will be changed by the same factor k. The response is simply scaled up or down in proportion.

As stated in the second paragraph of this section, certain non-linear relationships can be approximated by linear equations. The procedure is referred to as **linearisation** or **small perturbation analysis**. For a relationship such as $y = ax^2$ or $y = b\sqrt{x}$, Fig. 1.3,

Fig. 1.3 Illustrations of linearisation

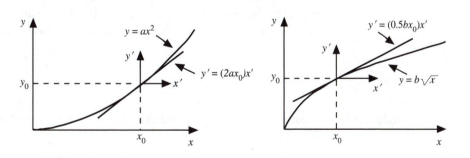

provided that deviation from a datum point (x_0, y_0) is small then the non-linear curve can be closely approximated by the tangent at this datum. If the variables are now defined relative to the datum values then the non-linear equation can be replaced by the linear equation $y' = kx'$, where $k = (dy/dx)$ at the datum, and x' and y' are relative to the datum. For the two relationships above, k would be $2ax_0$ and $0.5bx_0$ respectively.

Any given system component with a single input and single output will be represented by an nth order differential equation relating output to input. For the purposes of analysis and design the control engineer will use this in one of two ways. If 'classical control theory' is used, the equation will be transformed from the time domain into the Laplace domain by changing the independent variable from time t to the Laplace variable s, and then writing the differential equation in the form of a 'transfer function'. If 'modern control theory', developed in the 1960s, is used, additional variables will be introduced so that the nth order differential equation can be written as n first order differential equations, and these will be written in a shorthand matrix form. The form of the resultant 'state space model' will be illustrated later, but state space analysis will not be described since (a) there is insufficient space, (b) it provides less physical insight for the person new to the subject than the classical approach, and (c) its main use is for complex systems, particularly multi-input multi-output systems, and the first priority should be the understanding of simple systems.

1.3 Laplace transforms and transfer functions

The Laplace transform technique is widely used by the control engineer because it is a very convenient way of manipulating and solving differential equations. **Laplace transformation** is a process of changing the independent variable from t (time) to a complex variable $s = \sigma + j\omega$ (referred to as the Laplace operator). Following a commonly used convention, a function of time such as $f(t)$ or $x(t)$ will be written in lower case, and its Laplace transform $F(s)$ or $X(s)$ in capitals. The symbol $\mathscr{L}$ will be used to denote 'the Laplace transform of'. The **Laplace transform** of a function is defined by the equation:

$$F(s) = \mathscr{L}[f(t)] = \int_0^\infty e^{-st} f(t) dt \qquad [1.1]$$

with $f(t)$ assumed to be zero for $t < 0$.

Note that, although in the study of dynamics the variables are often implicitly assumed to be functions of time (i.e. x, $\dot{x}$, $\ddot{x}$), in this chapter they are explicitly shown as $x(t)$, $\dot{x}(t)$ or $dx(t)/dt$, etc., in order to highlight the distinction with the equivalent functions of s.

Example 1.1   Evaluate the Laplace transforms of the following functions of time $f(t)$ for $t \geq 0$, with $f(t) = 0$ for $t < 0$ in all cases:

(i) 1   (ii) t (iii) e^{-at} (iv) $1 + 5e^{-at}$ (v) $df(t)/dt$

Solution   Using the definition, Eq. 1.1:

$$\text{(i) } F(s) = \int_0^\infty e^{-st} dt = \left[\frac{e^{-st}}{-s}\right]_0^\infty = \frac{1}{s}[e^{-st}]_\infty^0 = \frac{1}{s}$$

$$\text{(ii)} \quad F(s) = \int_0^\infty t e^{-st} dt$$

$$= \left[\frac{t e^{-st}}{-s} \right]_0^\infty - \int_0^\infty \left(\frac{e^{-st}}{-s} \right) dt, \quad \text{by method of integration by parts}$$

$$= \int_0^\infty \left(\frac{e^{-st}}{s} \right) dt = \left[\frac{e^{-st}}{-s^2} \right]_0^\infty = \frac{1}{s^2}$$

$$\text{(iii)} \quad F(s) = \int_0^\infty e^{-at} e^{-st} dt = \int_0^\infty e^{-(s+a)t} dt = \left[\frac{e^{-(s+a)t}}{-(s+a)} \right]_0^\infty = \frac{1}{s+a}$$

$$\text{(iv)} \quad F(s) = \int_0^\infty (e^{-st} + 5e^{-(s+a)t}) dt = \frac{1}{s} + \frac{5}{s+a}$$

$$\text{(v)} \quad F(s) = \int_0^\infty \frac{df(t)}{dt} e^{-st} dt$$

$$= [e^{-st} f(t)]_0^\infty - \int_0^\infty f(t)(-se^{-st} dt) \ldots \text{integration by parts}$$

$$= -f(0) + s \int_0^\infty f(t) e^{-st} dt = sF(s) - f(0)$$

It should be appreciated that it is not necessary to evaluate Laplace transforms every time that Laplace transformation is to be carried out, since once evaluated they can be tabulated for future use. Table 1.1 lists those Laplace transform pairs which are used most in control system analysis; those in the upper section are so simple and widely used that they should be memorised, whilst those below can in general be looked up when required. Part (iv) of Example 1.1 illustrates two simple rules which show how to handle multiplication or division by a constant, and addition and subtraction of time

Table 1.1
Important Laplace transform pairs

$f(t)$	$F(s)$	$f(t)$	$F(s)$
1 (unit step)	$\dfrac{1}{s}$	$\dfrac{d}{dt}(f(t))$	$sF(s) - f(0)$
t (unit ramp)	$\dfrac{1}{s^2}$	$\dfrac{d^2}{dt^2}(f(t))$	$s^2F(s) - sf(0) - f^1(0)$
δt (unit impulse)	1	$\dfrac{d^n}{dt^n}(f(t))$	$s^nF(s) - s^{n-1}f(0) - s^{n-2}f^1(0) - \ldots f^{n-1}(0)$
e^{-at}	$\dfrac{1}{s+a}$		where $f^n(0) = \dfrac{d^n f(0)}{dt^n}$
$\sin \omega t$	$\dfrac{\omega}{s^2 + \omega^2}$	$e^{-at} \sin \omega t$	$\dfrac{\omega}{(s+a)^2 + \omega^2}$
$\cos \omega t$	$\dfrac{s}{s^2 + \omega^2}$	$e^{-at} \cos \omega t$	$\dfrac{(s+a)}{(s+a)^2 + \omega^2}$
		$\dfrac{\omega_n}{\sqrt{1 - \zeta^2}} e^{-\zeta \omega_n t} \sin \omega_n \sqrt{1 - \zeta^2} t$	$\dfrac{\omega_n^2}{s^2 + 2\zeta\omega_n s + \omega_n^2}$

functions, namely that $\mathscr{L}[Kf(t)] = KF(s)$, where K is a constant, and that $\mathscr{L}[f_1(t) \pm f_2(t)] = F_1(s) \pm F_2(s)$. Both can be seen to be true by looking at the definition Eq. 1.1.

Part (v) of the same example shows that the Laplace transform of the first derivative of $f(t)$ comprises two terms, $sF(s)$ minus the value of $f(t)$ when $t = 0$. Table 1.1 shows that the Laplace transform of the second derivative of $f(t)$ is $s^2F(s)$ minus two terms which are functions of the initial values of $f(t)$ and its first derivative, and that the Laplace transform of the nth derivative is $s^nF(s)$ minus n initial value terms. If all initial conditions are zero then Laplace transformation can be effected by the simple procedure of replacing d/dt by s and $f(t)$ by $F(s)$. Thus, if all initial conditions are zero, the Laplace transform of, say,

$$\frac{\mathrm{d}^4 f(t)}{\mathrm{d}t^4} + 5\frac{\mathrm{d}^3 f(t)}{\mathrm{d}t^3} + 2\frac{\mathrm{d}^2 f(t)}{\mathrm{d}t^2} + \frac{\mathrm{d}f(t)}{\mathrm{d}t} + 6f(t)$$

can be written immediately as

$$s^4 F(s) + 5s^3 F(s) + 2s^2 F(s) + sF(s) + 6F(s)$$

This can now be treated as an algebraic expression, and hence can be written as

$$(s^4 + 5s^3 + 2s^2 + s + 6)F(s)$$

Use of d/dt rather than dot notation makes transformation easier in the early stages of learning.

The Laplace transform technique for the solution of a differential equation relating an output $y(t)$ to an input $x(t)$ is a three-stage process. The first stage is transformation from the time domain to the s domain using the approach described above. The Laplace transform $X(s)$ of the input function $x(t)$ is then inserted, and the resultant equation treated as an algebraic equation and manipulated to find the value of $Y(s)$, the Laplace transform of the output, in terms of s. Finally, this expression for $Y(s)$ is then inverted back to the time domain, to yield the expression $y(t)$, by seeking an equivalent Laplace transform pair in a table of Laplace transforms, or if too complex by separating it by the partial fraction approach into the summation of simpler terms which can be found in the table.

Example 1.2   Find the functions of time $f(t)$ corresponding to the following functions $F(s)$:

$$\text{(i)} \quad \frac{1}{s+5} \qquad \text{(ii)} \quad \frac{1}{1+5s} \qquad \text{(iii)} \quad \frac{1}{s(s+1)(s+5)} \qquad \text{(iv)} \quad \frac{10}{s^2(1+0.5s)}$$

$$\text{(v)} \quad \frac{1}{s^2+4s+16} \qquad \text{(vi)} \quad \frac{s}{s^2+4s+20}$$

Solution   If an extensive table of Laplace transform pairs is available then the Laplace inverses can be written down for all of these by inspection. With only Table 1.1 available some prior manipulation is needed.

$$\text{(i)} \ F(s) = \frac{1}{s+5} \quad \therefore \ f(t) = \mathrm{e}^{-5t}$$

(ii) $F(s) = \dfrac{1}{1+5s} = \dfrac{0.2}{s+0.2}$ $\quad \therefore \ f(t) = 0.2e^{-0.2t}$

(iii) $F(s) = \dfrac{1}{s(s+1)(s+5)} = \dfrac{A}{s} + \dfrac{B}{s+1} + \dfrac{C}{s+5}$, say

$\quad \therefore \ 1 = A(s+1)(s+5) + Bs(s+5) + Cs(s+1)$

To evaluate A let $s = 0$ $\quad \therefore \ 1 = 5A$ $\quad \therefore \ A = 0.2$
To evaluate B let $s = -1$ $\quad \therefore \ 1 = -B(4)$ $\quad \therefore \ B = -0.25$
To evaluate C let $s = -5$ $\quad \therefore \ 1 = -5C(-4)$ $\quad \therefore \ C = 0.05$

$\quad \therefore \ F(s) = \dfrac{1}{s(s+1)(s+5)} = \dfrac{0.2}{s} - \dfrac{0.25}{s+1} + \dfrac{0.05}{s+5}$

$\quad \therefore \ f(t) = 0.2 - 0.25e^{-t} + 0.05e^{-5t}$

(iv) $F(s) = \dfrac{10}{s^2(1+0.5s)} = \dfrac{20}{s^2(s+2)} = \dfrac{A}{s^2} + \dfrac{B}{s} + \dfrac{C}{s+2}$, say

$\quad \therefore \ 20 = A(s+2) + Bs(s+2) + Cs^2$

To evaluate A let $s = 0$ $\quad \therefore \ 20 = 2A$ $\quad \therefore \ A = 10$

To evaluate C let $s = -2$ $\quad \therefore \ 20 = 4C$ $\quad \therefore \ C = 5$

To evaluate B insert the values for A and C in the above equation, and solve:

$20 = 10(s+2) + Bs(s+2) + 5s^2$

$\quad \therefore \ 20 = (B+5)s^2 + (2B+10)s + 20 \quad \therefore \ B = -5$

$\quad \therefore \ F(s) = \dfrac{10}{s^2(1+0.5s)} = \dfrac{10}{s^2} - \dfrac{5}{s} + \dfrac{5}{s+2}$

$\quad \therefore \ f(t) = 10t - 5 + 5e^{-2t}$

(v) $F(s) = \dfrac{1}{s^2 + 4s + 16} = \dfrac{(4)^2/16}{s^2 + 2(0.5)(4)s + (4)^2}$

$\quad \therefore \ f(t) = \dfrac{4}{16\sqrt{1-(0.5)^2}} e^{-(0.5)(4)t} \sin 4\sqrt{1-(0.5)^2}\,t = 0.289e^{-2t} \sin 3.46t$

(vi) $F(s) = \dfrac{s}{s^2 + 4s + 20} = \dfrac{s}{(s+2)^2 + 4^2}$

$\quad = \dfrac{s+2}{(s+2)^2 + 4^2} - \dfrac{0.5(4)}{(s+2)^2 + 4^2}$

$\quad \therefore \ f(t) = e^{-2t} \cos 4t - 0.5e^{-2t} \sin 4t$

$\quad = e^{-2t}(\cos 4t - 0.5 \sin 4t)$

$\quad = 1.118e^{-2t} \cos(4t + 0.464)$

Mathematics textbooks describe and prove numerous rules of Laplace transform analysis. Two very straightforward ones have already been illustrated in Example 1.1(iv). Memorising one further rule suffices for the early stages of control systems analysis—it is referred to as the **final value theorem**:

$$\lim_{t \to 0} f(t) = \lim_{s \to 0} sF(s) \qquad\qquad [1.2]$$

This enables the final value of $f(t)$, i.e. the value as $t \rightarrow \infty$, to be evaluated directly from a knowledge of $F(s)$. It is useful both directly and as a means of checking that a time solution $f(t)$ is not obviously in error.

Example 1.3   Find the final values of the time functions of Example 1.2

Solution

(i) $[f(t)]_{t=\infty} = \lim_{s \to 0} \dfrac{s}{s+5} = 0$

(ii) $[f(t)]_{t=\infty} = \lim_{s \to 0} \dfrac{s}{1+5s} = 0$

(iii) $[f(t)]_{t=\infty} = \lim_{s \to 0} \dfrac{1}{(s+1)(s+5)} = \dfrac{1}{5}$

(iv) $[f(t)]_{t=\infty} = \lim_{s \to 0} \dfrac{10}{s(1+0.5s)} = \infty$

(v) $[f(t)]_{t=\infty} = \lim_{s \to 0} \dfrac{s}{s^2 + 4s + 16} = 0$

(vi) $[f(t)]_{t=\infty} = \lim_{s \to 0} \dfrac{s^2}{s^2 + 4s + 20} = 0$

These agree with the results of letting $t = \infty$ in the expressions evaluated previously for $f(t)$.

The analytical techniques of classical control theory utilise the Laplace domain equivalent of the differential equation to describe the dynamic characteristics of a system. The **transfer function**, conventionally given the symbol $G(s)$, is defined as the ratio of the Laplace transform of the output to the Laplace transform of the input when all initial conditions are zero. Hence if the input and output functions are respectively $f_i(t)$ and $f_o(t)$ then

$$G(s) = \frac{\mathcal{L}[f_o(t)]}{\mathcal{L}[f_i(t)]} = \frac{F_o(s)}{F_i(s)} \tag{1.3}$$

when all initial conditions are zero. The initial conditions are not included since the dynamic characteristics of a system are independent of the initial state of the system. As illustration consider a dynamic system with input forcing function $x(t)$ and output $y(t)$ described by the second order differential equation

$$a\ddot{y}(t) + b\dot{y}(t) + cy(t) = x(t)$$

or

$$a\frac{d^2 y(t)}{dt^2} + b\frac{dy(t)}{dt} + cy(t) = x(t) \tag{1.4}$$

This transforms to

$$as^2 Y(s) + bsY(s) + cY(s) = X(s)$$

which can be treated as an algebraic equation, and hence can be written as

$$(as^2 + bs + c)Y(s) = X(s)$$

which can be rearranged to yield the transfer function

$$G(s) = \frac{Y(s)}{X(s)} = \frac{1}{as^2 + bs + c}$$ [1.5]

Equations 1.4 and 1.5 are alternative forms of mathematical model for the system.

1.4 Derivation of transfer functions

The procedure for deriving, by theoretical means, the transfer function for any system (and from it the differential equation, if required) broadly involves the following steps:

(a) Seek a qualitative understanding of how the system functions.
(b) Decide which variables could be used to describe the state of the system, assign symbols to them, and decide which are the input and output.
(c) Make certain simplifying assumptions.
(d) Decide what basic physical laws relate the variables and write down a set of equations; the number of equations should be one less than the number of variables.
(e) Laplace transform the equations, and combine them into a single equation by eliminating all variables except input and output.
(f) Rearrange to yield the ratio $\mathscr{L}$[output]/$\mathscr{L}$[input], which is the transfer function.

This procedure will be illustrated and explained by means of a series of examples in which transfer functions are derived for idealised representations of different types of physical process. The first few are mechanical systems incorporating masses constrained by springs and dampers, as would be studied in the subject of dynamics. Whenever a mass is free to move under the influence of applied forces then the physical equation which defines its motion is Newton's second law. Where springs of various forms are present (coil spring, leaf spring, deflected beam, twisted shaft, etc.), the physical equation relating force and deflection is Hooke's law, i.e. force is proportional to deflection from the free datum position. In such systems there is commonly also frictional resistance to movement. This may arise from viscous drag, and if so the force is proportional to the velocity of one part relative to another, as with a viscous damper used in vehicle suspension systems. If friction is constant, as with dry friction, then the equation is non-linear and analysis is beyond the scope of this book. Examples 1.4 and 1.5 derive the transfer function for a simple commonly occurring form of mechanical system, a single mass restrained by a spring and a damper. More complex mechanical systems such as those of Examples 1.6 and 1.7 are not significantly more difficult to tackle, though the arithmetic becomes laborious and great care must be taken to avoid calculation errors.

Example 1.4   Derive the transfer function for the mass–spring–damper arrangement shown in Fig. 1.4.

Solution   The mass m can move in a horizontal straight line, constrained by friction free guide surfaces, and is connected to a fixed vertical surface by a spring and a damper in parallel. In the absence of external force the mass will be at rest at a datum position with the spring at its free unstrained length. When a positive force $f(t)$ is applied to the mass it will move to the right, and as it does so the spring will compress and exert a resisting force, and the damper will also exert a force opposing the motion. If the applied force is

Fig. 1.4 Mass–spring–damper system

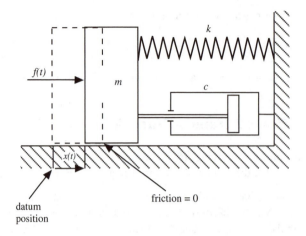

negative the mass will move to the left, opposed by tensile spring force and damper resistance. In general the applied force $f(t)$ can vary with time (as implied by the parameter t) and the resultant movement can be defined by $x(t)$, the position of the mass relative to its datum position. The input to the system is $f(t)$ and the output is $x(t)$, and since $x(t)$ fully defines the state of the system no other variables need be defined. (Those who have studied dynamics should recognise this as a single degree of freedom system.) Assume:

(a) the spring is ideal, i.e. force is proportional to deflection, and mass is negligible
(b) the damper is ideal, i.e. resisting force is proportional to velocity
(c) movement is not so great that the spring compresses to the solid condition, or that the damper reaches the limits of its range of movement.

Consider the forces acting on the mass at time t:

The externally applied force is $f(t)$.
The opposing spring force is $kx(t)$, where $k=$ spring stiffness, with units of N/m, say.
The opposing damper force is $c[dx(t)/dt]$, where $c=$ viscous damping coefficient, with units of Ns/m, say.

The net force applied to the mass is thus $f(t) - kx(t) - c[dx(t)/dt]$, and movement of the mass is governed by Newton's second law:

$$f(t) - kx(t) - c\frac{dx(t)}{dt} = m\frac{d^2x(t)}{dt^2}$$

or

$$m\frac{d^2x(t)}{dt^2} + c\frac{dx(t)}{dt} + kx(t) = f(t) \qquad [1]$$

This is the governing differential equation, and it is of second order. Transforming this into the Laplace domain, by replacing d/dt by s, $f(t)$ by $F(s)$ and $x(t)$ by $X(s)$ this can be written directly as

$$ms^2X(s) + csX(s) + kX(s) = F(s)$$

This can be treated as an algebraic equation and thus can be written as

$$(ms^2 + cs + k)X(s) = F(s)$$

Rearranging to obtain the ratio $\mathscr{L}$(output)$/\mathscr{L}$(input) the transfer function is found to be

$$G(s) = \frac{X(s)}{F(s)} = \frac{1}{ms^2 + cs + k} \qquad [2]$$

Example 1.5

Derive the transfer function for the same mass–spring–damper system, but with the direction of movement being vertical as shown in Fig. 1.5.

Fig. 1.5   Vertical mass–spring–damper system

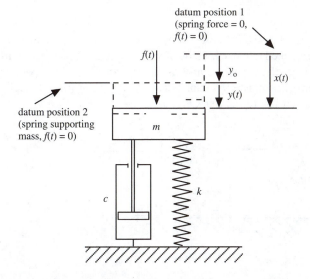

Solution   When the arrangement is vertical then an additional force acts on the mass, the gravitational attraction, or weight.

Let datum 1 represent the position of the mass when it is at rest and the spring force is zero, which requires $f(t)$ to be an upward force which just supports the weight. Let $x(t)$ be the absolute position of the mass relative to this datum.

In the absence of force $f(t)$ the mass will compress the spring by an amount $y_0 = mg/k$. Let this define a second datum, datum 2, and let $y(t)$ be the position of the mass relative to this datum.

Newton's second law thus takes the form

$$f(t) + mg - kx(t) - c\frac{\mathrm{d}x(t)}{\mathrm{d}t} = m\frac{\mathrm{d}^2x(t)}{\mathrm{d}t^2}$$

$$\therefore\; m\frac{\mathrm{d}^2x(t)}{\mathrm{d}t^2} + c\frac{\mathrm{d}x(t)}{\mathrm{d}t} + kx(t) - mg = f(t)$$

Now

$$x(t) = y_0 + y(t), \quad \therefore\; \frac{\mathrm{d}x(t)}{\mathrm{d}t} = \frac{\mathrm{d}y(t)}{\mathrm{d}t} \quad \text{and} \quad \frac{\mathrm{d}^2x(t)}{\mathrm{d}t^2} = \frac{\mathrm{d}^2y(t)}{\mathrm{d}t^2}$$

Substituting in the above differential equation gives

$$m\frac{\mathrm{d}^2y(t)}{\mathrm{d}t^2} + c\frac{\mathrm{d}y(t)}{\mathrm{d}t} + ky(t) = f(t)$$

The Laplace transform of this is

$$ms^2 Y(s) + csY(s) + kY(s) = F(s)$$

Hence the transfer function can be written as

$$G(s) = \frac{Y(s)}{F(s)} = \frac{1}{ms^2 + cs + k}$$

This is the same as when the arrangement is horizontal, from which it can be seen that the effect of gravity is simply a change of datum for the measurement of position.

Example 1.6   Figure 1.6 shows a mechanical system with two masses m_1 and m_2 which are constrained to move horizontally in a straight line by friction free guide surfaces, two springs of stiffness k_1 and k_2, and a damper with viscous damping coefficient c. Derive the transfer function relating the position $x(t)$ of the left end of spring k_1 to the position $y(t)$ of the mass m_2.

Fig. 1.6 Translational mechanical system

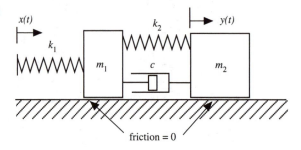

Solution   Increase in x will result in compression of spring k_1 causing movement of mass m_1, which in turn will cause movement of mass m_2 through the forces in the spring k_2 and the damper.

Assume springs are ideal (massless), damper movement is free of the end stops, and neither spring is compressed solid. The position of m_1 is neither $x(t)$ nor $y(t)$; call it $z(t)$. m_1 has three forces acting on it, and m_2 has two forces, as shown by the free body diagrams, Fig. 1.7. For each mass write down Newton's second law, i.e. net force = mass × acceleration:

$$k_1(x(t) - z(t)) - k_2(z(t) - y(t)) - c\left(\frac{dz(t)}{dt} - \frac{dy(t)}{dt}\right) = m_1 \frac{d^2 z(t)}{dt^2}$$

$$k_2(z(t) - y(t)) + c\left(\frac{dz(t)}{dt} - \frac{dy(t)}{dt}\right) = m_2 \frac{d^2 y(t)}{dt^2}$$

Fig. 1.7 Free body diagram derived from Fig. 1.6

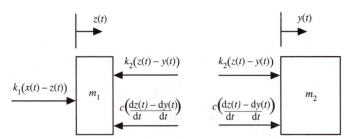

Transform into the Laplace domain:

$$k_1(X(s) - Z(s)) - k_2(Z(s) - Y(s)) - c(sZ(s) - sY(s)) = m_1s^2Z(s)$$
$$k_2(Z(s) - Y(s)) + c(sZ(s) - sY(s)) = m_2s^2Y(s)$$

Collect terms:

$$k_1X(s) = (m_1s^2 + cs + k_1 + k_2)Z(s) - (cs + k_2)Y(s)$$
$$(cs + k_2)Z(s) = (m_2s^2 + cs + k_2)Y(s)$$

Evaluate $Z(s)$ from the second equation and insert into the first:

$$k_1X(s) = (m_1s^2 + cs + k_1 + k_2)(m_2s^2 + cs + k_2)Y(s)/(cs + k_2)$$
$$- (cs + k_2)Y(s)$$

$$\therefore \ k_1(cs + k_2)X(s) = \{(m_1s^2 + cs + k_1 + k_2)(m_2s^2 + cs + k_2)$$
$$- (cs + k_2)^2\}Y(s)$$

$$\therefore \ G(s) = \frac{Y(s)}{X(s)} = \frac{k_1(cs + k_2)}{(m_1s^2 + cs + k_1 + k_2)(m_2s^2 + cs + k_2) - (cs + k_2)^2}$$

$$\therefore \ G(s) = \frac{Y(s)}{X(s)}$$

$$= \frac{k_1(cs + k_2)}{m_1m_2s^4 + c(m_1 + m_2)s^3 + (m_1k_2 + m_2k_2 + m_2k_1)s^2 + k_1cs + k_1k_2}$$

This is a fourth order transfer function, and corresponds to the differential equation (using dot notation as shorthand, e.g. $\dddot{y} = d^3y(t)/dt^3$):

$$m_1m_2\ddddot{y} + c(m_1 + m_2)\dddot{y} + (m_1k_2 + m_2k_2 + m_2k_1)\ddot{y} + k_1c\dot{y} + k_1k_2y$$
$$= k_1c\dot{x} + k_1k_2x$$

As with the simpler system of Examples 1.4 and 1.5 the effect of turning the system so that movement is in a vertical direction would be simply a change in the datum (at rest) positions of the masses. The springs will compress by m_2g/k_2 and $(m_1 + m_2)g/k_1$ respectively, and the equations of motion relative to these static equilibrium datum positions are exactly as above.

A brief discussion of the assumptions in these three examples will help to explain their significance. If the damper mass is not negligible then each moving part can be lumped with the mass to which it is attached, since it is subjected to the same acceleration. The spring mass, however, is distributed along its length which would lead to a partial differential equation; a reasonable approximation may be to consider the spring mass as though half were concentrated at each end. Clearly if movement is so large that a physical stop is reached on spring or damper then the equations of motion written down earlier do not remain valid. If motion is other than linear then additional positional variables and equations of motion must be included, and it is a much more complex situation. Frictional forces, unless viscous, introduce non-linearity.

Example 1.7   In a torsional shaft arrangement two cylindrical masses of inertias J_1 and J_2 are connected by a shaft of stiffness k_2 and a sleeve introducing viscous damping with coefficient c (Fig. 1.8). Mass J_1 is driven by a shaft of stiffness k_1. Determine the transfer function $\theta_o(s)/\theta_i(s)$.

Fig. 1.8 Rotational mechanical system

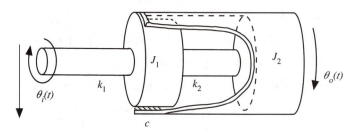

Solution   This is the rotational equivalent of the translational system of Example 1.6. Let $\theta_z(t)$ be the position of J_1. The free body diagrams are as shown in Fig. 1.9. The equations of motion are

$$k_1(\theta_i(t) - \theta_z(t)) - k_2(\theta_z(t) - \theta_o(t)) - c\left(\frac{d\theta_z(t)}{dt} - \frac{d\theta_o(t)}{dt}\right)$$

$$= J_1\frac{d^2\theta_z(t)}{dt^2}$$

$$k_2(\theta_z(t) - \theta_o(t)) + c\left(\frac{d\theta_z(t)}{dt} - \frac{d\theta_o(t)}{dt}\right) = J_2\frac{d^2\theta_o(t)}{dt^2}$$

and the transfer function is

$$G(s) = \frac{\theta_o(s)}{\theta_i(s)}$$

$$= \frac{k_1(cs + k_2)}{J_1J_2s^4 + c(J_1 + J_2)s^3 + (J_1k_2 + J_2k_2 + J_2k_1)s^2 + k_1cs + k_1k_2}$$

Fig. 1.9 Free body diagram derived from Fig. 1.8

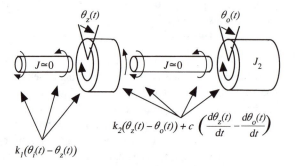

For electrical circuits the system variables are generally voltages and currents at various points in the circuit, and the basic physical laws defining the relationship between them are Ohm's law for resistors, the equivalent expressions for inductors and capacitors, Kirchoff's laws, etc. The components are discrete and the component behaviour is very close to the ideal.

Example 1.8   Derive the transfer function of the RLC circuit shown in Fig. 1.10 with input the voltage $v_i(t)$ and output the voltage $v_o(t)$.

Fig. 1.10   RLC circuit

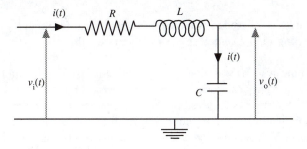

Solution   Assume that the input impedance is zero, i.e. that $v_i(t)$ is unaffected by the current drawn, and that the output impedance is infinite, i.e. whatever is connected at the output does not load the circuit so there is no current flow to the right.

Let $i(t)$ be the current through the resistor R and inductor L. Since the output impedance is assumed infinite $i(t)$ must also flow to the capacitor C. Let the voltage between R and L be $v(t)$.

$$v_i(t) - v(t) = i(t)R$$

$$v(t) - v_o(t) = L\frac{di(t)}{dt}$$

$$v_o(t) = \frac{1}{C}\int i(t)dt \quad \left(\text{or} \quad i(t) = C\frac{dv_o(t)}{dt}\right)$$

Laplace transforming these equations:

$$V_i(s) - V(s) = I(s)R \tag{1}$$

$$V(s) - V_o(s) = LsI(s) \tag{2}$$

$$V_o(s) = I(s)/Cs \quad (\text{or} \quad I(s) = CsV_o(s)) \tag{3}$$

There are 3 equations and 4 variables, hence eliminate $I(s)$ and $V(s)$. From [1] and [2]:

$$V_i(s) - V_o(s) = (R + Ls)I(s)$$

$$\therefore \quad V_i(s) - V_o(s) = (R + Ls)CsV_o(s), \quad \text{from [3]}$$

$$\therefore \quad V_i(s) = (LCs^2 + RCs + 1)V_o(s)$$

$$\therefore \quad \frac{V_o(s)}{V_i(s)} = \frac{1}{LCs^2 + RCs + 1} = \frac{1/LC}{s^2 + (R/L)s + 1/LC}$$

This has the same form as the transfer function for the mass–spring–damper of Examples 1.4 and 1.5. Physically the two systems are very different, but mathematically they are the same and if the coefficients have the same numerical values the dynamic behaviour would be identical. In dynamics this is referred to as a single degree of freedom system, in control as a **second order system**, or second order system component. It is generally arranged in a standard form, with a unity coefficient for s^2, as

$$\frac{Y(s)}{X(s)} = \frac{K\omega_n^2}{s^2 + 2\zeta\omega_n s + \omega_n^2}$$ [1.6]

in which ω_n is referred to as the **undamped natural frequency**, ζ as the **damping ratio** or **damping factor**, and K as the **steady state gain**.

For fluid systems with flow through piping from a supply and between various tanks the relevant physical relationships are flow continuity equations and an equation relating flow rate q to pressure drop Δp across a restriction. For a capillary type restrictor the Poiseuille equation shows that $q \propto \Delta p$. For an orifice type restrictor $q \propto \sqrt{\Delta p}$ which is non-linear and for linear analysis this is approximated by the tangent at the operating point and so by $q \propto \Delta p$, as was explained in Section 1.2.

Example 1.9   Water flows into a tank of cross sectional area A at rate $q_i(t)$ and out to atmosphere through a restrictor at rate $q_o(t)$, Fig. 1.11. Determine the transfer function relating the liquid depth $h(t)$ to the input flow rate.

Fig. 1.11   Fluid system

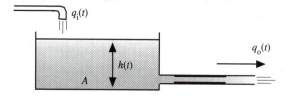

Solution   Net flow rate into tank = rate of change of volume in tank

$$\therefore \; q_i(t) - q_o(t) = A\frac{dh(t)}{dt}$$ [1]

Also flow rate out is proportional to the head of fluid

$$\therefore \; q_o(t) = Ch(t)$$ [2]

Laplace transforming Eqs [1] and [2] and eliminating $Q_o(s)$:

$$Q_i(s) - CH(s) = AsH(s)$$

$$\therefore \; \frac{H(s)}{Q_i(s)} = \frac{1}{C + As} = \frac{1/C}{1 + (A/C)s} = \frac{K}{1 + \tau s}$$

where $\tau = A/C$ and $K = 1/C$

For thermal systems the relevant equations are ones of energy conservation, heat transfer, temperature rise of a heated mass of fluid, etc.

Example 1.10   A volume of liquid, mass m and specific heat c, in a tank is heated by a heating coil and loses some heat to the surroundings, Fig. 1.12. What is the transfer function relating the temperature $T(t)$ to the rate of heat flow into the tank?

Fig. 1.12   Thermal
system

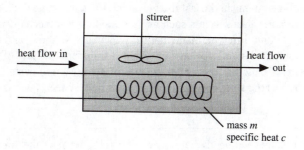

Solution   Let the rates of heat flow in and out be $q_i(t)$ and $q_o(t)$. The net rate of flow in determines
the rate of change of temperature:

$$\therefore \quad q_i(t) - q_o(t) = cm\frac{dT(t)}{dt}$$

and $q_o(t) \propto T(t) = kT(t)$ where $k =$ constant, and $T(t)$ is relative to ambient. Laplace
transforming, and eliminating $Q_o(s)$:

$$Q_i(s) - kT(s) = cmsT(s)$$

$$\therefore \quad \frac{T(s)}{Q_i(s)} = \frac{1}{k + cms} = \frac{1/k}{1 + (cm/k)s} = \frac{K}{1 + \tau s}$$

where $\tau = cm/k$ and $K = 1/k$.

The transfer function is as for Example 1.9, and the physical analogy is obvious.

Example 1.11   Derive the transfer function of the RC network shown in Fig. 1.13.

Fig. 1.13 RC network

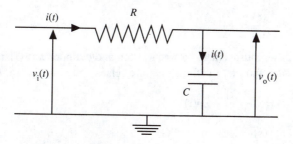

Solution   The assumptions are as for Example 1.8.

$$V_i(t) - V_o(t) = i(t)R$$

$$\text{and} \quad V_o(t) = \frac{1}{C}\int i(t)dt$$

Laplace transforming, and eliminating $I(s)$:

$$V_i(s) - V_o(s) = sCV_o(s)R$$

$$\therefore \quad \frac{V_o(s)}{V_i(s)} = \frac{1}{1 + RCs} = \frac{K}{1 + \tau s}$$

where $\tau = RC$ and $K = 1$

Each of the systems in the last three examples has the same form of transfer function, as would certain other systems such as a hydraulic servomechanism, and hence although physically quite different, mathematically they are the same, and they would exhibit the same form of dynamic behaviour. Such a transfer function of the form $b/(s+a)$ or $K/(1+\tau s)$ is called a **first order system** or a **simple lag system** and the value of τ is referred to as the **time constant** whilst K, as will be seen later, is the **steady state gain**.

Example 1.12   For a single acting hydraulic ram Fig.1.14, what is the transfer function relating piston position to input flow rate? Assume the fluid to be incompressible.

Fig. 1.14 Hydraulic ram

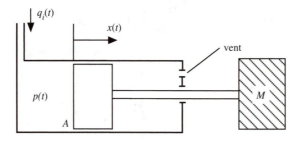

Solution   In the absence of any leakage past the piston, for flow continuity:

$$q_i(t) = A\frac{dx(t)}{dt}$$

$$\therefore\ Q_i(s) = AsX_i(s)$$

$$\therefore\ \frac{X_i(s)}{Q_i(s)} = \frac{1}{As} \tag{1}$$

If there is viscous leakage between piston and cylinder walls, let this be $Lp(t)$, where L is a constant and $p(t)$ the fluid pressure relative to atmosphere. For flow continuity:

$$q_i(t) = A\frac{dx(t)}{dt} + Lp(t)$$

$$\therefore\ Q_i(s) = AsX(s) + LP(s) \tag{2}$$

The pressure depends on the load inertia, and the second equation, necessary to be able to eliminate $P(s)$, is Newton's second law:

$$p(t)A = M\frac{d^2x(t)}{dt^2}$$

$$\therefore\ P(s)A = Ms^2X(s) \tag{3}$$

Combining Eqs [2] and [3]

$$Q_i(s) = AsX(s) + LMs^2X(s)/A$$

$$\therefore\ \frac{X(s)}{Q_i(s)} = \frac{1}{As\left(1 + \dfrac{LM}{A^2}s\right)} \tag{4}$$

It should be noted that since s represents differentiation, $1/s$ represents integration. In the absence of leakage, the position $x(t)$ is $1/A$ times the integral of the flow rate to the present time. If the flow rate is constant, the piston will move at constant velocity. When leakage is significant then there is an additional simple lag term with time constant LM/A^2.

Example 1.13 How is the transfer function of Example 1.12 affected if the fluid has bulk modulus B instead of being assumed incompressible ($B = \infty$)?

Solution $B =$ change in pressure/change in volume per unit volume

$\therefore$ change in volume $=$ volume under pressure $\times$ change in pressure$/B$

Differentiating, and letting V be the volume of fluid at pressure $p(t)$:

$$\frac{dv}{dt} = \frac{V}{B}\frac{dp(t)}{dt}$$

Now flow rate in $=$ piston displacement rate $+$ leakage flow rate $+$ flow needed to make up for compression of fluid.

$$\therefore \quad q_i(t) = A\frac{dx(t)}{dt} + Lp(t) + \frac{V}{B}\frac{dp(t)}{dt}$$

$$\therefore \quad Q_i(s) = AsX(s) + LP(s) + VsP(s)/B \qquad [1]$$

also $P(s)A = Ms^2X(s)$, as before $\qquad [2]$

Eliminating $P(s)$ from Eqs [1] and [2], and rearranging, gives:

$$\frac{X(s)}{Q(s)} = \frac{1}{s\left(\dfrac{VM}{BA}s^2 + \dfrac{LM}{A}s + A\right)}$$

The transfer function is thus an integration term with a second order component. The coefficient of s^2 varies with V, i.e. with piston position, but for small changes from datum can be considered to be constant. If $B = \infty$ this reduces to Eq. [4] of Example 1.12, and if also $L = 0$ it reduces to Eq. [1] of Example 1.12.

Example 1.14 Determine the form of the transfer function for a d.c. servomotor based on the following parameters: armature current $i_a =$ constant, field voltage $v_f =$ input, rotor inertia $= J$, rotor damping coefficient $= B$, shaft angular position $\theta_o =$ output.

Solution Torque $T(t) \propto$ flux cutting $\propto v_f(t)i_a = Cv_f(t)$, say

$$\text{also, } T(t) - B\frac{d\theta_o(t)}{dt} = J\frac{d^2\theta_o(t)}{dt^2}$$

$$\therefore \quad CV_f(s) - Bs\theta_o(s) = Js^2\theta_o(s)$$

$$\therefore \quad \frac{\theta_o(s)}{V_f(s)} = \frac{C}{s(B + Js)} = \frac{C/B}{s(1 + (J/B)s)}$$

This is the same form of transfer function as the ram with leakage, i.e. an integral term with a simple lag. If the output is considered to be shaft velocity rather than position then the transfer function is

$$\frac{s\theta_o(s)}{V_f(s)} = \frac{C/B}{1+(J/B)s}$$

which is a simple lag, like Examples 1.9 to 1.11.

The examples above illustrate how the basic physical equations governing system behaviour are used to derive theoretically the overall output–input relationship written in the form of a transfer function. It will be seen in the next chapter how, if a component is available for testing, the transfer function can be determined experimentally. The examples also illustrate how different physical systems can have the same mathematical model, and how secondary and tertiary effects increase the order of the transfer function.

1.5 Block diagram reduction

To undertake analysis or design each signal must be identified by a symbol and each block represented by a mathematical model relating the output to the input. Since analysis is carried out largely in the Laplace domain, the signals are written as functions of s rather than t, and the blocks are described by transfer functions, as in Fig. 1.15. By

Fig. 1.15 Block diagram for positional servomechanism

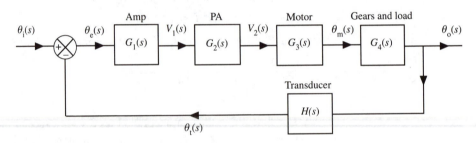

convention the symbol $G(s)$ is used for the transfer function of a block in the forward path of a feedback loop, and $H(s)$ for a block in a feedback path. Each transfer function would be derived either theoretically in the manner illustrated in the previous section, or experimentally by a method such as will be described in Section 2.6, or by a combination of the two.

A block diagram comprising several blocks, each with a relatively simple transfer function, can be reduced to a simpler form of diagram with fewer blocks, or to a single block with a more complex transfer function. The procedure is one of straightforward manipulation involving simple algebraic relationships. A number of blocks in series can be combined into a single block whose transfer function is the product of those for the blocks being combined. This can be illustrated by reference to the four blocks in the forward loop of Fig. 1.15, which can be combined into a single block to yield Fig. 1.16(a).

$$G(s) = \frac{\theta_o(s)}{\theta_e(s)} = \frac{\theta_o(s)}{\theta_m(s)}\frac{\theta_m(s)}{V_2(s)}\frac{V_2(s)}{V_1(s)}\frac{V_1(s)}{\theta_e(s)} = G_4(s)G_3(s)G_2(s)G_1(s)$$

$$= G_1(s)G_2(s)G_3(s)G_4(s) \qquad [1.7]$$

Fig. 1.16 Reduced forms of block diagram, corresponding to Fig. 1.15

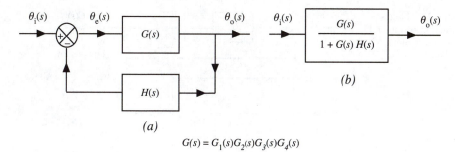

$$G(s) = G_1(s)G_2(s)G_3(s)G_4(s)$$

Also in Fig. 1.16a,

$$\theta_e(s) = \theta_i(s) - \theta_o(s)H(s) \qquad [1.8]$$

and

$$\theta_o(s) = \theta_e(s)G(s) \qquad [1.9]$$

Combining Eqs 1.8 and 1.9 to eliminate $\theta_e(s)$ gives

$$\theta_o(s) = \theta_i(s)G(s) - \theta_o(s)G(s)H(s)$$

$$\therefore \; \theta_o(s)(1 + G(s)H(s)) = \theta_i(s)G(s)$$

$$\therefore \; \frac{\theta_o(s)}{\theta_i(s)} = \frac{G(s)}{1 + G(s)H(s)} \qquad [1.10]$$

Thus the system of Fig. 1.15 can be reduced to a single block with transfer function as shown in Fig. 1.16(b). The equivalence of Fig. 1.16(a) and (b) given by Eq. 1.10 will be referred to frequently, and should be memorised, together with the simple derivation. From Eq. 1.10 could be written down the overall differential equation relating input $\theta_i(t)$ and output $\theta_o(t)$, so this process of block diagram reduction can be seen to be a straightforward method for combining the differential equations of the individual components and eliminating intermediate variables.

Where there are minor feedback loops the diagram can be reduced to a single block in stages, as illustrated by the following examples.

Example 1.15　Derive the overall transfer function for the system represented by the block diagram shown in Fig. 1.17.

Fig. 1.17 Block diagram, Example 1.15

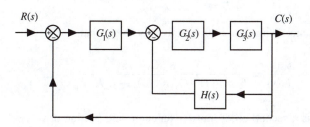

Solution　Using R, G_1, H... as shorthand for $R(s)$, $G_1(s)$, $H(s)$..., and Eq. 1.10 to eliminate the inner loop, an equivalent diagram is Fig. 1.18(a) which reduces first to Fig. 1.18(b), then to Fig. 1.18(c).

Fig. 1.18 Stages of
reduction of Fig. 1.17
to a single block

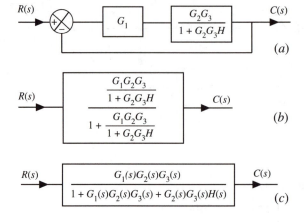

(a)

(b)

(c)

Example 1.16   Derive the transfer function relating the angular position (radians) to the input voltage (proportional to radians) of the servomechanism represented by the block diagram of Fig. 1.19.

Fig. 1.19 Block
diagram, Example 1.16

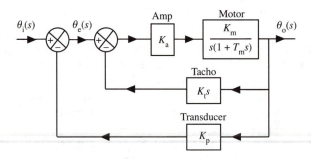

Solution   Either by the procedure of Example 1.15, or noting that the diagram is equivalent to Fig. 1.16(a) with

$$G(s) = \frac{K_a K_m}{s(1 + T_m s)}, \quad H(s) = K_t s + K_p$$

and hence using Eq. 1.10 directly it is found that

$$\frac{\theta_o(s)}{\theta_i(s)} = \frac{K_a K_m}{T_m s^2 + (1 + K_a K_m K_t)s + K_a K_m K_p}$$

This is a second order transfer function, and writing it as

$$\frac{\theta_o(s)}{\theta_i(s)} = \frac{\dfrac{1}{K_p} \cdot \dfrac{K_a K_m K_p}{T_m}}{s^2 + \left(\dfrac{1 + K_a K_m K_t}{T_m}\right)s + \dfrac{K_a K_m K_p}{T_m}}$$

it can be seen that the gain is $1/K_p$, $\omega_n = \sqrt{(K_aK_mK_p/T_m)}$, and the damping factor can be evaluated as $\zeta = 0.5(\sqrt{(1/K_aK_mK_pT_m)} + \sqrt{(K_aK_mK_t^2/K_pT_m)})$. The effect of the inclusion of the minor feedback loop incorporating tachogenerator is thus to increase ζ without change in gain or ω_n.

Example 1.17 Derive the overall transfer function for the system represented by Fig. 1.20.

Solution In this diagram two of the loops interlink. Eliminate by altering one of the loops to a non-interlinking equivalent, Fig. 1.21(a), and then reduce the diagram in steps as in Fig. 1.21(b) to (d).

Fig. 1.20 Block diagram, Example 1.17

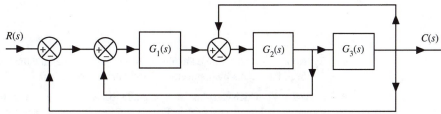

Fig. 1.21 Stages of reduction of Fig. 1.20 to a single block

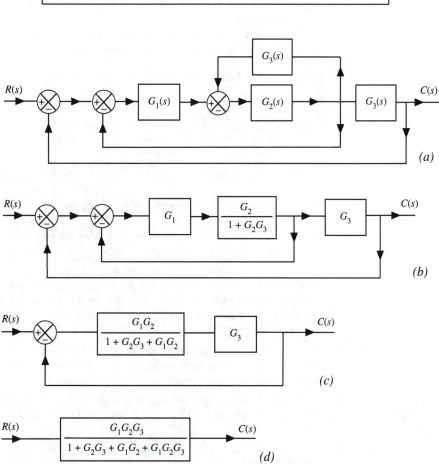

Problems

1 For each of the following feedback systems draw a qualitative block diagram to represent the system: (a) water level control in a tank with a ball float controlled valve to regulate the supply; (b) thermostatic control of a hot water room heating radiator; (c) the control loop for movement of a robot joint.

2 Starting from the defining relationship evaluate the Laplace transforms of the following functions of time: (a) $5t$; (b) $10+2t$; (c) $1+e^{-3t}$; (d) $f(t)=1$ for $0<t\le 10$ and zero for $t>10$.

3 Write down, with the aid of Table 1.1, the Laplace transforms of the following expressions assuming initial conditions are zero: (a) $\ddot{c}(t)+5\dot{c}(t)+c(t)$; (b) $e^{-2t}\cos 0.5t$; (c) $5+20t$; (d) $x(t)=10\dot{y}(t)+5y(t)$; (e) $c(t)=2\sin\omega t$.

4 Determine the Laplace inverse of each of the following expressions for $F(s)$:

(a) $\dfrac{3}{s+4}$ (b) $\dfrac{2}{s(s+3)}$ (c) $\dfrac{1}{(1+0.1s)(1+0.5s)}$ (d) $\dfrac{5}{s(s^2+6s+25)}$

(e) $\dfrac{5}{s^2(s^2+6s+25)}$

5 By letting $t\to\infty$ determine the final values for each of the time functions $f(t)$ derived in Problem 4. Check the results by application of the final value theorem.

6 A mass m kg which can slide along a guide rod within a frame is held in a central position by a spring on each side as shown in Fig. P1.1. The stiffness of each spring, which can exert either tensile or compressive force, is k N/m, and the friction between mass and guide rod is viscous with damping coefficient c Ns/m. Determine the transfer function relating the position of the mass to the force applied in an axial direction.

Fig. P1.1

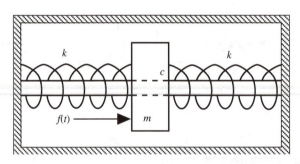

7 For the mechanical system shown in Fig. P1.2 derive the transfer function relating the position of the mass to the input, when the input is considered to be (a) the force $f(t)$, (b) the position $x(t)$.

8 An electric motor drives a pump by a relatively long shaft. Figure P1.3 shows schematically the two rotors of inertia J_1 and J_2, restrained by viscous damping with coefficients C_1 and C_2, and connected by the shaft of stiffness K. What are the transfer functions relating motor angular position $\theta_1(t)$ and pump angular positions $\theta_2(t)$ to the torque $T(t)$ which is applied to the rotor of the motor? What would the latter transfer function be if the shaft were infinitely stiff?

9 Consider the drive between the motor and pump of Problem 8 to be through a pair of gears with reduction ratio $1:n$ and rigid shafts, Fig. P1.4. Determine the transfer function relating motor shaft angle to the torque applied to the rotor of the motor.

10 Show that the transfer functions relating voltage $v_o(t)$ to $v_i(t)$ for the two electrical circuits of Fig. P1.5 have the form $K(1+\alpha Ts)/(1+Ts)$. What are the values of K, T and α in terms of R_1, R_2 and C?

11 Derive the transfer function for the electrical circuit shown in Fig. P1.6, where the input is the voltage $v_i(t)$ and the output the voltage $v_o(t)$.

Fig. P1.2

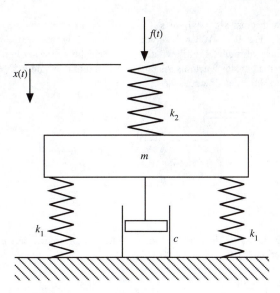

Fig. P1.3

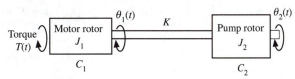

Fig. P1.4

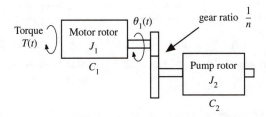

Fig. P1.5

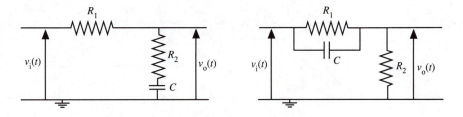

Fig. P1.6

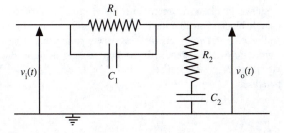

12   Two tanks with cross-sectional areas A_1 and A_2 are interconnected by a pipe which has linear restriction R_1, Fig. P1.7. Water flows into tank 1 at flow rate $q_i(t)$, and flows out from tank 2 through a linear restrictor R_2 at flow rate $q_o(t)$. The depths of water in the two tanks are $h_1(t)$ and $h_2(t)$. Derive the transfer functions relating (a) the output flow rate $q_o(t)$, (b) the depth $h_2(t)$, (c) the depth $h_1(t)$ to the input flow rate $q_i(t)$.

Fig. P1.7

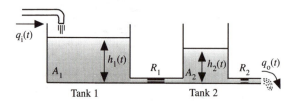

Tank 1   Tank 2

13   By reducing the block diagram of Fig. P1.8 to a single block derive the form of the overall transfer function $C(s)/R(s)$.

Fig. P1.8

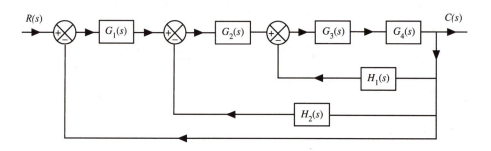

14   By reducing the block diagram of Fig. P1.9 to a single block derive the form of the overall transfer function $C(s)/R(s)$.

Fig. P1.9

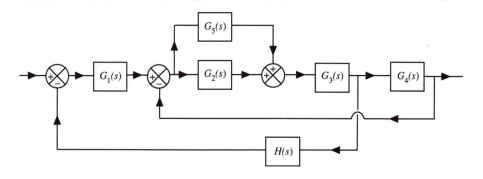

15   Figure P1.10 shows in block diagram form an electrical servomechanism. By reducing the diagram to a single block show that the overall transfer function is of second order. What are the values of overall gain, ζ and ω_n in terms of the system parameters? Calculate the values of gain, ζ and ω_n, and write down the differential equation relating position $\theta_m(t)$ to position command voltage $v_i(t)$ when $K_m = 0.8$ rad/sec/volt, $T_m = 1$ sec, $K_a = 10$, $K_p = 1.6$ volts/rad and $K_t = 0.4$ volts/rad/sec.

Fig. P1.10

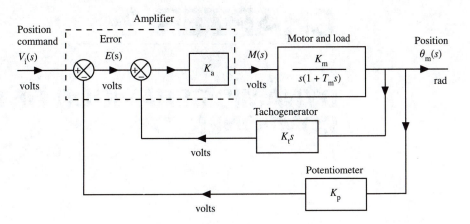

CHAPTER 2

DYNAMIC BEHAVIOUR OF SYSTEM COMPONENTS

In Chapter 1 it was shown how a variety of different dynamic system components can be represented by mathematical models in the form of block diagrams and associated transfer functions. The objective of this chapter is to explain the main methods available for investigating the dynamic behaviour of such components, and to build up understanding of how the behaviour is influenced by the nature and the parameters of the mathematical model. It considers transfer functions of first order, second order and higher orders which may represent either individual system components, or groups of components in series or in a loop, which have been reduced to a single overall transfer function by the block diagram reduction methods of Section 1.5. This should form a sound foundation for study, in Chapter 3, of the behaviour of feedback systems with particular reference to stability and the magnitude of the steady state error. The design of feedback systems with one or more loops to achieve acceptable dynamic behaviour can then be tackled with purpose and insight in Chapter 4.

2.1 Methods of analysis

In dynamics, when studying vibrations in mechanical systems two forms of excitation are considered – release from a non-equilibrium state (i.e. from an initial non-zero position and/or velocity) for the case of free vibrations, and a sinusoidal input (of position or force) for the case of forced vibrations. In general, and for systems in a wider sense, the forcing function, or input signal, to a system component or a complete system can have an infinite variety of forms and often cannot be described deterministically as an explicit function of time. To simplify analysis, and indeed to permit analysis without resort to computer solution, it is necessary to consider certain specific forms of forcing function which are mathematically easy to handle. It is also desirable to have simple forms of forcing function to facilitate the interpretation of results.

Probably the most useful and frequently used forcing function is the **step change of input** where the input is changed suddenly from one value to a new constant value, Fig. 2.1(a). It is mathematically simple and, just as important, occurs commonly in practice.

Fig. 2.1 Transient forcing functions: (*a*) step; (*b*) ramp; (*c*) impulse

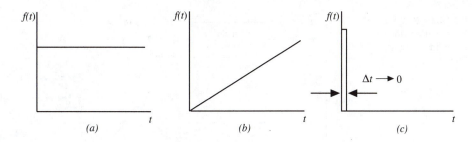

It is the most severe disturbance which can be applied, and knowledge of how the system output will be affected is of considerable interest. When a car accelerator is suddenly depressed or released, how does the engine speed react? When a thermostat setting is moved to a new value, how does the actual temperature change with time? When a valve controlling flow into a tank is suddenly opened by a certain amount, how is the liquid level in the tank affected? It should be noted that the release of a mechanical system from an initial condition of zero velocity and non-zero position is equivalent to a step change of input. With reference to Fig. 1.5 (p. 13) the sudden displacement upwards of the base by some distance d will result in an identical response $x(t)$ as would depression of the mass by a distance d followed by release from rest, the only difference being a change of datum. Another simple forcing function is the **ramp function**, or step change of derivative of input, Fig. 2.1(b) (a sudden change of velocity if the input is a position, or of rate of change of temperature if the input is a temperature, say). This is a less severe form of disturbance, relevant where step changes are not possible, or not normal. A third form of input function which is useful for analysis is an **impulse** Fig. 2.1(c), but in practice it is less valuable than the other two because to be large enough to produce an identifiable response, especially if there is some noise present, the amplitude of the impulse must be so high that the system may be damaged or driven into a region of non-linear operation. Step, ramp, and impulse functions are all **transient disturbances**; the system normally starts from a steady state and reattains a steady state condition after the effect of the disturbance has died out. In analysis it is the nature of the transient response leading to the new steady state and the error between the steady state and the ideal value which are studied.

A different form of forcing function which is also mathematically simple is a **sinusoidal input**. For a linear system (defined earlier in Section 1.2, and with further explanation in the last paragraph of this section) when a sine wave is applied at the input then the output signal will build up in some transient manner until a steady state is attained in which the output is also sinusoidal, of the same frequency as the input, but generally differing in amplitude and phase, Fig. 2.2. **Frequency** or **harmonic response analysis** ignores the transient part of the response and considers how the amplitude and the phase shift of the steady state output relative to the input (generally referred to in control analysis as the **magnitude** and **phase**) vary with the frequency of the sinusoidal input signal. Sinusoidal inputs have relevance since they commonly occur in practice, e.g. in mechanical systems undergoing vibratory movement and in a.c. electrical systems, and since, by Fourier decomposition, any periodic signal of arbitrary waveform can be broken down into the summation of a series of sine and cosine waves with specific amplitudes at the fundamental frequency and integer multiples of this frequency. A third class of forcing function is a **statistical signal** with a waveform which is wholly or partially random. The mathematics of analysis and the interpretation of results are less straightforward and will not be discussed beyond a brief outline in Section 4.4.

Fig. 2.2 Response of linear system to sinusoidal forcing function

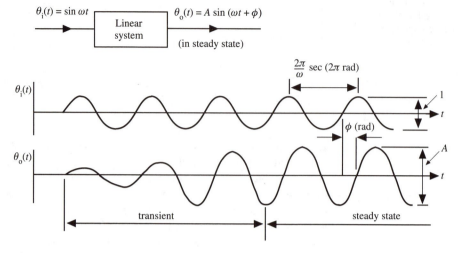

For step, ramp, and impulse inputs the process of analysis is one of solving the governing differential equations for the forcing functions of interest. The Laplace transform technique for obtaining such solutions was outlined in Section 1.3, and will be used in this chapter. The resultant output response may be presented as an analytical function of time, or simply as a plot of output against time. If the mathematical model is of high order, or if the response to a more complicated forcing function is required, then the engineer will often solve the equations by undertaking simulation on a computer to obtain the time response. Analogue computers, which were formerly used for this, have now been almost completely superseded by high level simulation languages on digital computers.[†]† The remainder of this chapter in part will build up an understanding of the general nature of transient response and of the forms of response to be expected for different types of transfer function.

For a sinusoidal input to a system of known transfer function, the output magnitude and phase for different frequency values can be determined analytically, graphically, or by using an appropriate computer program. Two general principles are involved. The magnitude and phase information is contained in the complex expression $G(j\omega)$, the transfer function $G(s)$ with s replaced by $j\omega$, often referred to as the **sinusoidal transfer function**. Justification for this can be found in the book by Schwarzenbach and Gill (see Bibliography), but since $s = \sigma + j\omega$ it will be appreciated that $G(j\omega)$ is a particular case of $G(s)$. Also, it should be apparent that if several blocks are in series (as in the forward loop of Fig. 1.2, p. 4), with the output of one being the input to the next, then the overall magnitude will be the product of the magnitudes for the individual blocks, and the overall phase will be the sum of the phases (i.e. phase shifts) for the individual blocks. If, as is commonly the case, the magnitudes are expressed in decibels or db (where a magnitude X has the value $20 \log_{10} X$ in decibels) then the overall magnitude is the sum of the individual magnitudes. If a higher order transfer function can be factorised then it is often easier, with less risk of arithmetic error, to determine magnitude and phase for the component parts and combine them rather than to determine the harmonic response

---

[†] Probably the most widely available software for control system analysis and design is MATLAB with its Control System Toolbox extension. With this a few lines of program code suffices for much of the analysis. Software called CODAS is available at low cost to accompany the book by Golten and Verwer (see Bibliography), and this is very convenient and easy to use.

information directly. The harmonic response information is normally presented on one of three types of plot (Fig. 2.3):

(a) Polar plot: an Argand diagram in which a vector representing the input is assumed to lie along the positive real axis, and for each value of frequency an output vector represents magnitude and phase (with lags represented by clockwise rotation). The harmonic response information is contained in the locus of the ends of the output vectors for frequency values ranging from zero to infinity.

(b) Bode plot: a pair of plots, of magnitude in decibels and phase in degrees on linear scales, plotted against frequency on a logarithmic scale. Two valuable features are that straight line approximations make the plots relatively easy to draw for first and second order components (as will be seen in Sections 2.3 and 2.4), and that obtaining overall magnitude and phase from curves for the constituent parts simply involves graphical addition. Change in gain implies a vertical shift of the magnitude curve, with no change to the phase curve.

(c) Nichols chart: a plot of magnitude (db) against phase (degrees). This plot, as will be seen in the next chapter, has particular relevance for feedback systems, since if the plot is drawn from the open loop information then closed loop harmonic response information can be read off directly.

The curves on each of these three plots show the same harmonic response information, and each highlights rather different facets and can be used for different additional purposes.

Fig. 2.3 Plots showing harmonic response information: (*a*) polar plot; (*b*) Bode plot; (*c*) Nichols chart

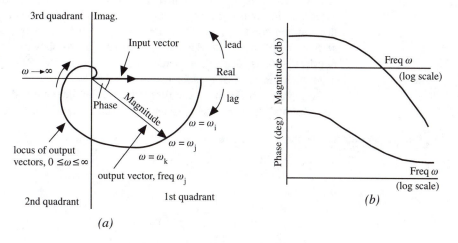

(*a*)

(*b*)

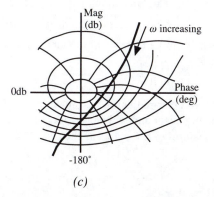

(*c*)

In the early stages of analysis, and wherever possible in the later stages also, system components are assumed to be linear. As was stated in Section 1.2 a linear system is one for which the principle of superposition is valid. This means that responses to individual forcing functions can be superposed, e.g. if the response to an input $x_1(t)$ is $y_1(t)$, and to $x_2(t)$ is $y_2(t)$ then the response to the combined input $x_1(t)+x_2(t)$ would be $y_1(t)+y_2(t)$. It also means that alteration of the amplitude of the input function by a factor k will result simply in a scaling of the output response by k, with no change of shape. Hence it is normal to consider step, ramp, impulse and sinusoidal forcing functions to be of unit magnitude – a **unit step** implies a change from zero to unity at time $t=0$, a **unit ramp** implies an input increasing at a rate of 1 unit per second from zero at $t=0$, a **unit impulse** is one of unit area, and a sinusoidal input has amplitude unity. For any other forcing function amplitude the output response is scaled accordingly. The differential equation of a linear system contains no products or powers of the variables or their derivatives, and for a time invariant system the coefficients are constant.

2.2 Response of second order system to transient inputs

To derive the response of a mass–spring–damper system to a step change of input the dynamicist would solve the governing second order differential equation by the classical methods of solution involving a particular integral and the complementary function. The control engineer would employ the Laplace transform technique to obtain the solution for this (and for other forms of system) in the manner shown in this section, which draws on Sections 1.3 to 1.5. It was shown in Section 1.4 that other physical systems such as the rotational equivalent and the RLC circuit of Example 1.8 have the same form of transfer function.

To represent a general second order system let the input and output variables be $x(t)$ and $y(t)$, and the coefficients arranged in such a way that the differential equation is written in the form

$$\ddot{y} + 2\zeta\omega_n\dot{y} + \omega_n^2 y = K\omega_n^2 x$$

or

$$\frac{d^2y(t)}{dt^2} + 2\zeta\omega_n\frac{dy(t)}{dt} + \omega_n^2 y(t) = K\omega_n^2 x(t) \qquad [2.1]$$

After Laplace transformation this becomes

$$s^2 Y(s) + 2\zeta\omega_n s Y(s) + \omega_n^2 Y(s) = K\omega_n^2 X(s)$$

which can be rearranged to yield the transfer function

$$\frac{Y(s)}{X(s)} = \frac{K\omega_n^2}{s^2 + 2\zeta\omega_n s + \omega_n^2} \qquad [2.2]$$

which appeared as Eq. 1.6 in Section 1.4. K, the steady state gain, is a simple multiplying factor and will be considered to have value unity for the time being. The Laplace transform of the output is then

$$Y(s) = \frac{\omega_n^2 X(s)}{s^2 + 2\zeta\omega_n s + \omega_n^2} \qquad [2.3]$$

Unit step input

For a unit step input, i.e. $x(t) = 1$ for $t \geq 0$ and $x(t) = 0$ for $t < 0$, it has been shown in Example 1.1 that $X(s) = 1/s$, hence the output is given by

$$Y(s) = \frac{\omega_n^2}{s(s^2 + 2\zeta\omega_n s + \omega_n^2)} \qquad [2.4]$$

The quadratic has roots $\lambda_1, \lambda_2 = -\zeta\omega_n \pm \omega_n\sqrt{(\zeta^2 - 1)}$. When the numerical values of the coefficients are such that the damping ratio $\zeta < 1$ then these roots will be a complex conjugate pair, when $\zeta = 1$ they will be real and equal, and when $\zeta > 1$ they will be real and unequal. If an available table of Laplace transform pairs contains the expression Eq. 2.4 then for $\zeta < 1$ the equivalent time function, the response of the system to a unit step, can be written down directly as

$$y(t) = 1 - \frac{e^{-\zeta\omega_n t}}{\sqrt{(1 - \zeta^2)}} \sin\left(\omega_n\sqrt{(1 - \zeta^2)}t + \cos^{-1}\zeta\right) \qquad [2.5]$$

For $\zeta > 1$ the quadratic can be factorised and Eq. 2.4 written as

$$Y(s) = \frac{\lambda_1\lambda_2}{s(s - \lambda_1)(s - \lambda_2)} \qquad [2.6]$$

If the expression in Eq. 2.6 or an equivalent form such as $1/s(1 + \tau_1 s)(1 + \tau_2 s)$ appears in the table then the response can, probably after some manipulation, be written down as

$$y(t) = 1 + \frac{\lambda_2}{\lambda_1 - \lambda_2}e^{\lambda_1 t} + \frac{\lambda_1}{\lambda_2 - \lambda_1}e^{\lambda_2 t} \qquad [2.7]$$

If, as with Table 1.1, these functions do not appear then $Y(s)$ must first be separated into partial fractions, as was done in Example 1.2, to yield simpler functions which can be inverted. Using a normalised time scale $\omega_n t$ the step response, Eqs 2.5 and 2.7, for a range of values of ζ is shown in Fig. 2.4(a). The smaller the value of ζ the more oscillatory is the response, and the 'frequency' of the oscillation is the damped natural frequency $\omega_d = \omega_n\sqrt{(1 - \zeta^2)}$. When $\zeta \geq 1$ there is no overshoot and for large values of ζ the response becomes very sluggish. A value of $\zeta \approx 0.7$ can be thought of as giving a 'good' response – small overshoot, quick settling. The steady state part of the response is unity; for $\zeta < 1$ the transient part is the second term of Eq. 2.5 which is an oscillation whose amplitude decays to zero as $t \to \infty$, and for $\zeta > 1$ the transient part is the pair of exponential terms of Eq. 2.7 which also decay to zero as $t \to \infty$ since λ_1 and λ_2 are negative. If the gain K has a value other than unity then $y(t)$ contains this as a multiplying factor and the response is scaled accordingly.

Unit ramp input

For a unit ramp input, i.e. $x(t) = t$ for $t \geq 0$ and $x(t) = 0$ for $t < 0$, $X(s) = 1/s^2$ and the output is

$$Y(s) = \frac{\omega_n^2}{s^2(s^2 + 2\zeta\omega_n s + \omega_n^2)} \qquad [2.8]$$

If this expression can be found in a table of Laplace transform pairs then for $\zeta < 1$ the response to a unit ramp can be written down directly as

$$y(t) = t - \frac{2\zeta}{\omega_n} + \frac{e^{-\zeta\omega_n t}}{\omega_n\sqrt{(1 - \zeta^2)}} \sin\left(\omega_n\sqrt{(1 - \zeta^2)}t + 2\cos^{-1}\zeta\right) \qquad [2.9]$$

Fig. 2.4 Response of unity gain second order system to: (*a*) unit step; (*b*) unit ramp; (*c*) unit impulse input

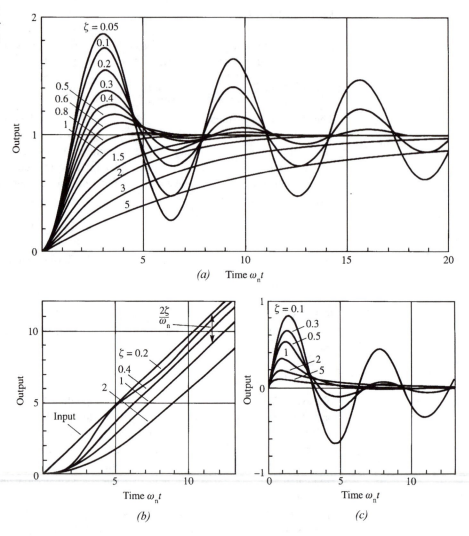

Figure 2.4(b) shows this response for values of $\zeta = 0.2$ and 0.4. It will be appreciated that each will be one of a family of curves with oscillations having the character of the curves in Fig. 2.4(a). For $\zeta \geq 1$ there is no oscillation, as typified by the curves for $\zeta = 1$ and 2 in Fig. 2.4(b), and

$$y(t) = t - \frac{2\zeta}{\omega_n} + A_1 e^{\lambda_1 t} + A_2 e^{\lambda_2 t} \qquad [2.10]$$

The steady state part of the response is $t - 2\zeta/\omega_n$, and hence there is a steady error of $-2\zeta/\omega_n$ between the input t and the output, once the transient part of the response (terms 3 and 4 where λ_1 and λ_2 are negative) has decayed to zero, Fig. 2.4(b). If the expression Eq. 2.8 and its factored equivalent are not available in Laplace transform tables then Eqs 2.9 and 2.10 can be found by carrying out partial fraction expansion prior to inversion, but this is tedious and prone to error with symbols rather than numerical values. Examples 2.1 and 2.2 illustrate.

Unit impulse

$X(s) = 1$, and hence

$$Y(s) = \frac{\omega_n^2}{s^2 + 2\zeta\omega_n s + \omega_n^2}$$

which, by reference to Table 1.1 can be seen to invert to

$$y(t) = \frac{\omega_n e^{-\zeta\omega_n t}}{\sqrt{(1 - \zeta^2)}} \sin (\omega_n \sqrt{(1 - \zeta^2)}t) \qquad \text{[2.11]}$$

For values of gain K other than unity inspection of the above analysis will show that the output response will simply be scaled by the value K.

Example 2.1   For a system component with transfer function $G(s) = Y(s)/X(s) = 1/s^2 + 5s + 4$ find the response $y(t)$ to an input forcing function $x(t)$ which is: (a) a unit step; (b) a unit ramp.

Solution $\omega_n = \sqrt{4} = 2$, $2\zeta\omega_n = 5$ $\therefore$ $\zeta = 1.25$
The quadratic thus has real roots and can be factorised into $(s + 1)(s + 4)$

(a) $X(s) = 1/s$

$$\therefore Y(s) = \frac{1}{s(s^2 + 5s + 4)} = \frac{1}{s(s + 1)(s + 4)} = \frac{A}{s} + \frac{B}{s + 1} + \frac{C}{s + 4}$$

$$\therefore 1 = A(s + 4)(s + 1) + Bs(s + 4) + Cs(s + 1)$$

$$\therefore A = 1/4, B = -1/3, C = 1/12$$

$$\therefore Y(s) = \frac{1/4}{s} - \frac{1/3}{s + 1} + \frac{1/12}{s + 4}$$

$$\therefore y(t) = 1/4 - 1/3e^{-t} + 1/12e^{-4t}$$

(b) $X(s) = 1/s^2$

$$\therefore Y(s) = \frac{1}{s^2(s^2 + 5s + 4)} = \frac{1}{s^2(s + 1)(s + 4)} = \frac{A}{s^2} + \frac{B}{s} + \frac{C}{s + 1} + \frac{D}{s + 4}$$

$$\therefore 1 = A(s + 1)(s + 4) + Bs(s + 1)(s + 4) + Cs^2(s + 4) + Ds^2(s + 1)$$

$$\therefore A = 1/4, \ B = -15/48, \ C = 1/3, \ D = -1/48$$

$$\therefore y(t) = 1/4t - 15/48 + 1/3e^{-t} - 1/48e^{-4t}$$

When $t = 0$, in both cases, $y(t) = 0$. When $t \to \infty$ $y(t)$ for a unit step tends to 1/4, which is consistent with a gain K of 1/4 when $G(s)$ is written in the standard form, Eq. 2.2. These are useful quick checks for obvious errors in the answers. With a ramp input the output rises at 1/4 of the rate of the input (consistent with $K = 1/4$), and the output in the steady state lags the ramp $t/4$ by an amount 15/48.

Example 2.2   For a system component with transfer function $G(s) = Y(s)/X(s) = 1/(s^2 + 2s + 4)$ find the response $y(t)$ to a forcing function $x(t)$ which is: (a) a step change of magnitude 10; (b) an input increasing at 2 units per second.

Solution
$$\omega_n = \sqrt{4} = 2, \quad 2\zeta\omega_n = 2 \quad \therefore \; \zeta = 0.5$$

The quadratic thus has complex roots and cannot be factorised to give real factors.

(a) $x(t) = 10 \quad \therefore \; X(s) = 10/s$

$$\therefore \; Y(s) = \frac{10}{s(s^2 + 2s + 4)} = \frac{A}{s} + \frac{Bs + C}{s^2 + 2s + 4}$$

$$10 = A(s^2 + 2s + 4) + (Bs + C)s$$

$$\therefore \; A = 2.5, \; B = -2.5, \; C = -5$$

$$\therefore \; Y(s) = \frac{2.5}{s} - \frac{2.5s + 5}{s^2 + 2s + 4}$$

$$= \frac{2.5}{s} - \frac{2.5s + 5}{(s + 1)^2 + (\sqrt{3})^2}$$

$$= \frac{2.5}{s} - \frac{2.5(s + 1)}{(s + 1)^2 + (\sqrt{3})^2} - \frac{(2.5/\sqrt{3})(\sqrt{3})}{(s + 1)^2 + (\sqrt{3})^2}$$

$$\therefore \; y(t) = 2.5 - 2.5e^{-t} \cos(\sqrt{3})t - (2.5/\sqrt{3})e^{-t} \sin(\sqrt{3})t$$

$$= 2.5 - 2.5e^{-t}(\cos 1.732t + 0.577 \sin 1.732t)$$

$$\therefore \; y(t) = 2.5 - 2.89e^{-t} \sin(1.732t + 1.047)$$

Alternatively, this can be derived by using Eq. 2.5, the Laplace inverse of Eq. 2.4, with a multiplying factor of $10/4 = 2.5$.

(b) $x(t) = 2t \quad \therefore \; X(s) = 2/s^2$

$$\therefore \; Y(s) = \frac{2}{s^2(s^2 + 2s + 4)} = \frac{A}{s^2} + \frac{B}{s} + \frac{Cs + D}{s^2 + 2s + 4}$$

$$\therefore \; 2 = A(s^2 + 2s + 4) + Bs(s^2 + 2s + 4) + (Cs + D)s^2$$

$$\therefore \; A = 0.5, \; B = -0.25, \; C = 0.25, \; D = 0$$

$$\therefore \; Y(s) = \frac{0.5}{s^2} - \frac{0.25}{s} + \frac{0.25s}{s^2 + 2s + 4}$$

$$= \frac{0.5}{s^2} - \frac{0.25}{s} + \frac{0.25(s + 1)}{(s + 1)^2 + (\sqrt{3})^2} - \frac{(0.25/\sqrt{3})\sqrt{3}}{(s + 1)^2 + (\sqrt{3})^2}$$

$$\therefore \; y(t) = 0.5t - 0.25 + 0.25e^{-t} \cos(\sqrt{3})t - (0.25/\sqrt{3})e^{-t} \sin(\sqrt{3})t$$

$$= 0.5t - 0.25 + 0.25e^{-t}(\cos 1.732t - 0.577 \sin 1.732t)$$

$$\therefore \; y(t) = 0.5t - 0.25 + 0.288e^{-t} \sin(1.732t + 2.094)$$

Alternatively, this can be derived by using Eq. 2.9.

2.3 Response of second order system to sinusoidal input

The response of a mass–spring–damper system to a sinusoidal input forcing function, which the dynamicist would derive by solution of the differential equation by classical methods, would be derived by the control engineers by following the approach outlined in Section 2.1, as will now be shown. The transfer function, Eq. 2.2, with K again considered to have value unity for the time being is

$$\frac{Y(s)}{X(s)} = G(s) = \frac{\omega_n^2}{s^2 + 2\zeta\omega_n s + \omega_n^2}$$

Replacing s by $j\omega$ yields $G(j\omega)$ which can then be rationalised to give the coordinates on the polar plot:

$$G(j\omega) = \frac{\omega_n^2}{(\omega_n^2 - \omega^2) + j2\zeta\omega\omega_n} = \frac{1}{\left(1 - \left(\frac{\omega}{\omega_n}\right)^2\right) + j2\zeta\left(\frac{\omega}{\omega_n}\right)}$$

$$= \frac{\omega_n^2[(\omega_n^2 - \omega^2) - j2\zeta\omega\omega_n]}{(\omega_n^2 - \omega^2)^2 + (2\zeta\omega\omega_n)^2} = \frac{1 - \left(\frac{\omega}{\omega_n}\right)^2 - j2\zeta\frac{\omega}{\omega_n}}{\left(1 - \left(\frac{\omega}{\omega_n}\right)^2\right)^2 + \left(2\zeta\frac{\omega}{\omega_n}\right)^2}$$

The real and imaginary parts are both functions of the forcing frequency ω, and for a given value of ω_n there is a family of curves for various values of ζ, Fig. 2.5(a). For the Bode plot, Fig. 2.5(b), the magnitude is

$$-20 \log_{10}\sqrt{\left[\left(1 - \frac{\omega^2}{\omega_n^2}\right)^2 + \left(\frac{2\zeta\omega}{\omega_n}\right)^2\right]}\, \text{db}$$

and the phase is

$$-\tan^{-1}\left(\frac{2\zeta\dfrac{\omega}{\omega_n}}{1 - \dfrac{\omega^2}{\omega_n^2}}\right)$$

For $\omega/\omega_n \ll 1$ $\quad$ mag $\approx -20 \log_{10} 1 = 0$, phase $= 0°$

$$\omega/\omega_n \gg 1 \quad \text{mag} \approx -20 \log_{10}\left(\frac{\omega^2}{\omega_n^2}\right) = -40 \log_{10}\left(\frac{\omega}{\omega_n}\right),$$

$$\text{phase} \approx -\tan^{-1}\frac{\omega_n}{\omega} = -180°$$

$$\omega/\omega_n = 1 \quad \text{mag} = -20 \log_{10}\left(\frac{2\zeta\omega}{\omega_n}\right), \text{ phase} = -90°$$

On the magnitude plot the curves tend towards a pair of straight line asymptotes – zero db for very low frequencies, and a line of slope -40 db per decade of frequency (i.e. factor of 10 increase) intersecting the zero db line at the frequency ω_n. This intersection is referred to as the **break point** or **corner frequency**. For low values of ζ there is a relatively narrow peak of magnitude corresponding to resonance at a frequency ω close to the undamped natural frequency ω_n. The peak value tends to infinity as $\zeta \to 0$. On the phase plot it can be seen that for low values of ζ there is a very rapid increase of phase lag for small increases in frequency around ω_n from a value close to $0°$ to a value close to $180°$, whilst for large values of ζ the phase change takes place gradually over a wide range of frequency.

Fig. 2.5 Response of unity gain second order system to harmonic input: (*a*) polar plot; (*b*) Bode plot

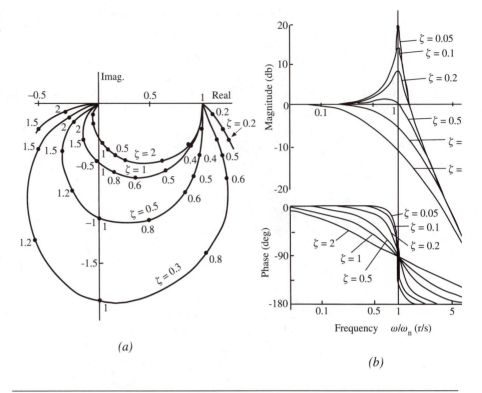

(*a*)

(*b*)

Example 2.3 Show on a polar plot and a Bode plot the frequency response of the system component of Example 2.2, i.e. $G(s) = 1/(s^2 + 2s + 4)$

Solution

$$G(j\omega) = \frac{1}{(4 - \omega^2) + j2\omega} = \frac{(4 - \omega^2) - j2\omega}{(4 - \omega^2)^2 + 4\omega^2}$$

$$= \left(\frac{4 - \omega^2}{\omega^4 - 4\omega^2 + 16}\right) - j\left(\frac{2\omega}{\omega^4 - 4\omega^2 + 16}\right)$$

By inserting values of ω, real and imaginary coordinates of points on the polar plot can be calculated, e.g.

ω	0	0.2	0.5	1	1.5	2	2.5	3	4	10
Real	0.25	0.25	0.249	0.231	0.145	0	−0.075	−0.082	−0.058	−0.010
Imag.	0	−0.025	−0.066	−0.154	−0.249	−0.25	−0.166	−0.098	−0.038	−0.002

Alternatively, polar coordinates can be calculated:

$$|G(j\omega)| = \frac{\sqrt{(4 - \omega^2)^2 + 4\omega^2}}{(4 - \omega^2)^2 + 4\omega^2} = \frac{1}{\sqrt{(\omega^4 - 4\omega^2 + 16)}}$$

$$\underline{/G(j\omega)} = -\tan^{-1}\frac{2\omega}{4 - \omega^2}$$

ω	0	0.2	0.5	1	1.5	2	2.5	3	4	10	100
Mag	0.25	0.251	0.258	0.277	0.288	0.250	0.182	0.128	0.069	0.010	0.0001
db	−12.04	−12.00	−11.78	−11.14	−10.81	−12.04	−14.78	−17.85	−23.18	−39.82	−80.00
Phase	−0	−5.8	−14.9	−33.7	−59.7	−90.0	−65.8	−50.2	−146.3	−168.2	−178.9

Figure 2.6(a) shows this frequency response information on a polar plot. This is a second order component with $\omega_n = \sqrt{4} = 2$ r/s, and $2\zeta\omega_n = 2$ $\therefore$ $\zeta = 0.5$. The straight

Fig. 2.6 Polar plot and Bode plot, Example 2.3

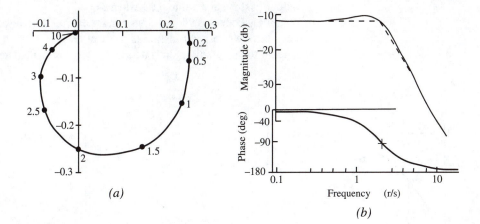

(a)

(b)

line approximation for magnitude is -12 db up to the frequency 2 r/s, then decreases at 40 db/decade for frequencies beyond this. The true curve is obtained by transferring the curve for $\zeta = 0.5$ from Fig. 2.5(b) onto this plot (centred on $\omega_n = 2$ r/s), Fig. 2.6(b). The phase is $0°$ for very low frequencies, $-180°$ for very high frequencies, $-90°$ for $\omega = 2$ r/s, and varies with frequency as the phase curve for $\zeta = 0.5$ in Fig. 2.5(b), which is transferred to Fig. 2.6(b).

2.4 Response of first order system

First order components, of which Examples 1.9 to 1.11 are typical, are the simplest dynamic elements appearing in a control loop and occur very commonly, but have no equivalent in mechanical systems involving springs, masses, and damping elements. The transient response for simple inputs is very easily obtained by the Laplace transform approach. The transfer function, corresponding to the differential equation $\tau \dot{y}(t) + y(t) = Kx(t)$ is

$$\frac{Y(s)}{X(s)} = \frac{K}{1 + \tau s}$$

The output, with K taken as unity for the time being, is

$$Y(s) = \frac{X(s)}{1 + \tau s}$$

Unit step

For a unit step, $X(s) = 1/s$

$$\therefore \; Y(s) = \frac{1}{s(1 + \tau s)} = \frac{1/\tau}{s(s + 1/\tau)} = \frac{1}{s} - \frac{1}{s + 1/\tau}$$

$$\therefore \; y(t) = 1 - e^{-t/\tau} \qquad\qquad [2.12]$$

This response is shown in Fig. 2.7(a). The following characteristics should be noted:

(i) the steady state value is 1, the transient component is $-e^{-t/\tau}$

(ii) the output has reached value 0.63 when $t = \tau$

(iii) the output has values 0.95 and 0.98 when $t = 3\tau$ and 4τ respectively

(iv) the tangent to $y(t)$ at $t = 0$ intersects $y = 1$ at $t = \tau$

(v) the tangent to $y(t)$ at any point intersects $y = 1$ τ seconds later.

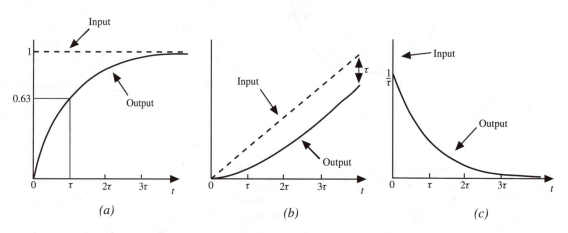

Fig. 2.7   Response of unity gain first order system to: (*a*) unit step; (*b*) unit ramp; (*c*) unit impulse input

Unit ramp

For a unit ramp, $X(s) = 1/s^2$

$$\therefore\ Y(s) = \frac{1}{s^2(1 + \tau s)} = \frac{1/\tau}{s^2(s + 1/\tau)} = \frac{A}{s^2} + \frac{B}{s} + \frac{C}{s + 1/\tau}$$

$$\therefore\ 1/\tau = As(s + 1/\tau) + B(s + 1/\tau) + Cs^2$$

$$\therefore\ A = 1,\ B = -\tau,\ C = \tau$$

$$\therefore\ y(t) = t - \tau + \tau e^{-t/\tau} \tag{2.13}$$

This response is shown in Fig. 2.7(b). The steady state response is $t - \tau$, thus there is a steady state error of value $-\tau$. The transient component of the response is $\tau e^{-t/\tau}$ which decays to 2% of τ by time $t = 4\tau$.

Unit impulse

For a unit impulse, $X(s) = 1$

$$\therefore\ Y(s) = \frac{1}{1 + \tau s} = \frac{1/\tau}{s + 1/\tau}$$

$$\therefore\ y(t) = 1/\tau e^{-t/\tau} \tag{2.14}$$

The steady state response is zero, the transient decaying from an initial value of $1/\tau$ to 2% of $1/\tau$ in time 4τ, Fig. 2.7(c).

Sinusoidal input

For a sinusoidal input

$$G(j\omega) = \frac{1}{1+j\omega\tau} = \frac{1-j\omega\tau}{1+\omega^2\tau^2} = \frac{1}{1+\omega^2\tau^2} - j\left(\frac{\omega\tau}{1+\omega^2\tau^2}\right)$$

$$= \frac{1}{\sqrt{(1+\omega^2\tau^2)}} \underline{/\tan^{-1}\omega\tau}$$

The polar plot, Fig. 2.8(a), is a semicircle with centre (0.5, j0) and radius 0.5. For very small ω the magnitude is unity and the phase is zero. As $\omega \to \infty$ the magnitude decreases and tends to zero while the phase is a lag which tends to a maximum of 90°. The lag is 45° when $\omega = 1/\tau$.

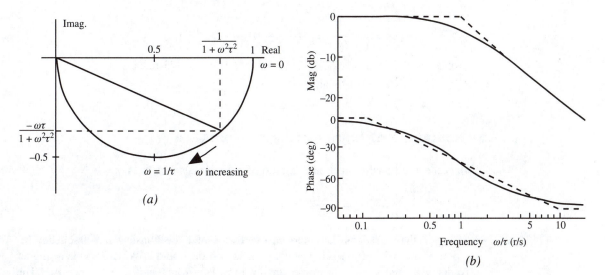

Fig. 2.8 Response of unity gain first order system to harmonic input: (*a*) polar plot; (*b*) Bode plot

The Bode plot, Fig. 2.8(b), can be represented by straight line approximations: Magnitude $= -20\log_{10}\sqrt{(1+\omega^2\tau^2)}$. Phase $= -\tan^{-1}\omega\tau$. For

$$\omega\tau \ll 1, \text{ i.e. } \omega \ll 1/\tau \, |G(j\omega)| \approx -20\log_{10} 1 = 0, \underline{/G(j\omega)} \approx 0°$$

$$\omega\tau = 1, \text{ i.e. } \omega = 1/\tau \, |G(j\omega)| = -20\log_{10}\sqrt{2} = -3\,db, \underline{/G(j\omega)} = -45°$$

$$\omega\tau \gg 1, \text{ i.e. } \omega \gg 1/\tau \, |G(j\omega)| \approx -20\log_{10}\omega\tau, \underline{/G(j\omega)} \approx -90°$$

The harmonic response is approximated on the magnitude plot by two lines intersecting at the break point or corner frequency $\omega = 1/\tau$: a zero db line, and a line of slope -20 db/decade of frequency passing through 0 db at $\omega = 1/\tau$. The true curve is 3 db lower at $\omega = 1/\tau$, 1 db lower at $\omega = 0.5(1/\tau)$ and $2(1/\tau)$, and 0.04 db lower at $0.1(1/\tau)$ and $10(1/\tau)$.

On the phase plot phase is approximated by the three lines 0° to the frequency $0.1(1/\tau)$, $-90°$ for $\omega \geq 10(1/\tau)$, and a linear variation between (passing through

$-45°$ for $\omega = 1/\tau$). The true curve is S-shaped with maximum error of $5.5°$ at $\omega \approx$ $0.1(1/\tau)$, $0.5(1/\tau)$, $2(1/\tau)$ and $10(1/\tau)$ as shown in Fig. 2.8(b).

Example 2.4   A first order component has unity gain and a time constant of 5 seconds. What is the output value 10 seconds after application of a unit step change of input? What is the steady state output for an input forcing function of $\sin t$?

Solution

$$G(s) = \frac{Y(s)}{X(s)} = \frac{1}{1 + 5s} = \frac{0.2}{s + 0.2}$$

When $X(s) = 1/s$, a unit step

$$Y(s) = \frac{0.2}{s(s + 0.2)} = \frac{1}{s} - \frac{1}{s + 0.2}$$

$$\therefore \ y(t) = 1 - e^{-0.2t}$$

For $t = 10$ $y(10) = 1 - e^{-2} = 0.865$

$$G(j\omega) = \frac{1}{1 + 5j\omega} = \frac{1}{\sqrt{(1 + 25\omega^2)}} \ \underline{/-\tan^{-1}5\omega}$$

For $\omega = 1$ $|G(j\omega)| = 0.196$, $\underline{/G(j\omega)} = -78.7° = -1.373$ radians

$$\therefore \ y(t) = 0.196 \sin (t - 1.373)$$

2.5 Response of higher order systems

In general a transfer function is a ratio of polynomials in s, thus

$$\frac{Y(s)}{X(s)} = G(s) = \frac{P_n(s)}{P_d(s)}$$

The last three sections have derived expressions for the transient and frequency response where the denominator polynomial is of order 1 and 2. When $P_d(s)$ is of higher order n then it can be factorised in the form $(s-p_1)(s-p_2)(s-p_3)\ldots(s-p_n)$. The parameters $p_1, p_2, p_3 \ldots p_n$ are referred to as the **poles of the transfer function**, and are values which if substituted for s would make it infinite. The equation $P_d(s) = 0$, obtained by equating to zero the denominator of the transfer function relating output to input of a system, is defined as the **characteristic equation** of the system (it is called this because, as will be seen shortly, its roots determine the character of the response of the system). $p_1, p_2, p_3 \ldots p_n$ are therefore also the **roots of the characteristic equation**. (Note that the terms poles and roots are often used interchangeably, though poles tends to be used when referring to components and roots when referring to an overall system). The transfer function can then be written as

$$\frac{Y(s)}{X(s)} = \frac{P_n(s)}{(s - p_1)(s - p_2)(s - p_3)\ldots(s - p_n)}$$

For a unit step input

$$Y(s) = \frac{P_n(s)}{s(s - p_1)(s - p_2)(s - p_3)\ldots(s - p_n)}$$

which can be separated by partial fraction expansion to give

$$Y(s) = \frac{A_0}{s} + \frac{A_1}{s - p_1} + \frac{A_2}{s - p_2} + \frac{A_3}{s - p_3} + \dots \frac{A_n}{s - p_n}$$

where $A_0, A_1, A_2 \dots A_n$ are constants. Laplace inversion yields the time response

$$y(t) = A_0 + A_1 e^{p_1 t} + A_2 e^{p_2 t} + A_3 e^{p_3 t} + \dots A_n e^{p_n t}$$

For a unit ramp input

$$Y(s) = \frac{P_n(s)}{s^2(s - p_1)(s - p_2)(s - p_3) \dots (s - p_n)}$$

which after partial fraction expansion and Laplace inversion gives

$$y(t) = B' t + B_0 + B_1 e^{p_1 t} + B_2 e^{p_2 t} + B_3 e^{p_3 t} + \dots B_n e^{p_n t}$$

where $B', B_0, B_1, B_2 \dots B_n$ are constants which are not the same as $A_0, A_1, A_2 \dots A_n$

For a unit impulse input

$$Y(s) = \frac{P_n(s)}{(s - p_1)(s - p_2)(s - p_3) \dots (s - p_n)}$$

and

$$y(t) = C_1 e^{p_1 t} + C_2 e^{p_2 t} + C_3 e^{p_3 t} + \dots C_n e^{p_n t}$$

where $C_1, C_2, C_3, \dots C_n$ is a further set of constants.

Reference back to Sections 2.4 and 2.2 will show that the previously obtained responses for first and second order systems are as above with $n = 1$ and $n = 2$ respectively.

The response to a transient forcing function can thus be seen in all cases to be made up of a steady state component (A_0, $B' t + B_0$, zero for step, ramp, and impulse respectively) together with a transient component $\sum_{i=1}^{n} A_i e^{p_i t}$. The form of contribution to the response of each term of the summation depends on the value of the pole or root p_i, and its magnitude on the coefficient A_i. Each root p_i can lie anywhere in the complex s plane. If p_i lies on the negative real axis, i.e. is real and negative, $A_i e^{p_i t}$ will have value A_i at $t = 0$ and will decay to zero at a rate which depends on the numerical value of p_i. If p_i lies on the positive real axis then $A_i e^{p_i t}$ will increase exponentially without bound. If p_i is on the positive imaginary axis then one of the other roots must be its conjugate and the two terms together result in a cosine wave of frequency given by the distance from the origin. If p_i is complex, then one of the other roots must be its conjugate and the pair of terms results in an oscillation whose amplitude is decaying or increasing exponentially, depending on whether the roots have negative or positive real part. This is illustrated by Fig. 2.9. For complex roots, $-\zeta\omega_n \pm j\omega_n\sqrt{1 - \zeta^2}$, the value of ω_n is the distance from the origin, and the damping ratio ζ the cosine of the angle relative to the origin, as shown in Fig. 2.9(b).† A number of important factors must be understood. If any of the roots lies in the half plane to the right of the imaginary axis then there will be a component in the response which is increasing without bound, i.e. the output does not settle to a steady state, and the system is said to be unstable. The contribution to the response of poles to the left of, but close to, the imaginary axis are terms which decay to zero more slowly than those from poles farther to the left. The poles nearest to the origin thus dominate the response, and the contributions of poles to the left become less and less the farther they are from the origin. The coefficient A_i determines the weighting of each factor, and generally poles which are well to the left and thus decay quickly will also have small coefficient values. The roots/poles nearest

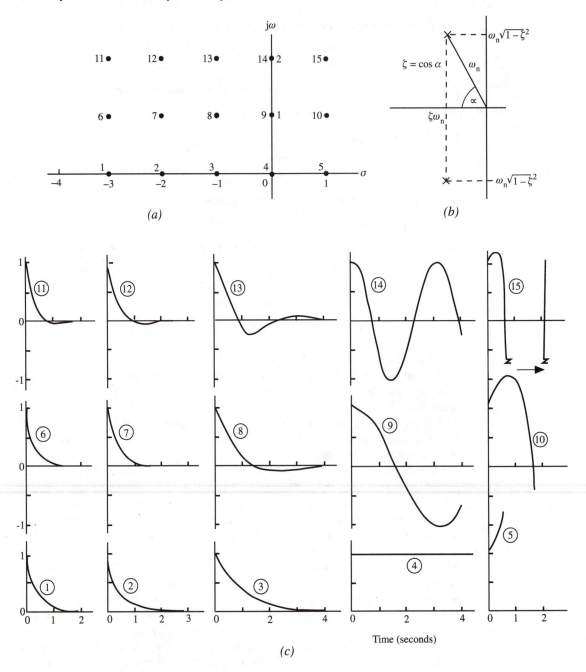

Fig. 2.9 Effect of root position on contribution to transient response: (*a*) root positions in *s* plane; (*b*) value of ω_n and ζ corresponding to a pair of complex conjugate roots; (*c*) resultant change to initial value of +1 for root positions 1 to 15, where traces 6–15 result from conjugate pairs of roots

the origin are thus referred to as the **dominant roots** or poles, and they dominate the response. Secondary factors, when included in the transfer function as in Example 1.13, will result in poles more distant than the dominant poles. These ideas are best appreciated with the aid of numerical examples.

Example 2.5   A system has transfer function $1/(1+s)(1+2s)(1+5s)$. Find the response of the system to: (a) a unit step input; (b) a unit ramp input. For the step input what is the time taken to reach within 2% of the steady value?

Solution   Let input and output be $\theta_i(t)$ and $\theta_o(t)$ respectively, or in Laplace notation $\theta_i(s)$ and $\theta_o(s)$. Now,

$$\frac{\theta_o(s)}{\theta_i(s)} = \frac{1}{(1+s)(1+2s)(1+5s)} = \frac{0.1}{(s+1)(s+0.5)(s+0.2)}$$

(a) $\theta_i(s) = 1/s$

$$\therefore \; \theta_o(s) = \frac{0.1}{s(s+1)(s+0.5)(s+0.2)} = \frac{A}{s} + \frac{B}{s+1} + \frac{C}{s+0.5} + \frac{D}{s+0.2}, \; \text{say.}$$

Now,

$$A = 0.1/(1)(0.5)(0.2) = 1$$

$$B = 0.1/(-1)(-0.5)(-0.8) = -0.25$$

$$C = 0.1/(-0.5)(0.5)(-0.3) = 1.333$$

$$D = 0.1/(-0.2)(0.8)(0.3) = -2.083$$

$$\therefore \; \theta_o(s) = \frac{1}{s} - \frac{0.25}{s+1} + \frac{1.333}{s+0.5} - \frac{2.083}{s+0.2}$$

$$\therefore \; \theta_o(t) = 1 - 0.25e^{-t} + 1.333e^{-0.5t} - 2.083e^{-0.2t} \qquad [1]$$

As a check that there is no obvious error Eq. [1] gives $\theta_o(0) = 0$. The root at $s = -0.2$ dominates in that the transient decays the most slowly, and the coefficient -2.083 is the largest of the three. Terms 2 and 3 will decay more rapidly than term 4, hence the settling time is given by

$$1 - 2.083e^{-0.2t} = 0.98$$

i.e.

$$2.083 = 0.02e^{0.2t}$$

$$\therefore \; t = 23.2 \text{ sec}$$

This is close to four times the dominant time constant.

(b) $\theta_i(s) = 1/s^2$

$$\therefore \; \theta_o(s) = \frac{0.1}{s^2(s+1)(s+0.5)(s+0.2)} = \frac{A}{s^2} + \frac{B}{s} + \frac{C}{s+1} + \frac{D}{s+0.5} + \frac{E}{s+0.2}$$

$$= \frac{1}{s^2} - \frac{8}{s} + \frac{0.25}{s+1} - \frac{2.667}{s+0.5} + \frac{10.417}{s+0.2}$$

$$\therefore \; \theta_o(t) = t - 8 + 0.25e^{-t} - 2.667e^{-0.5t} + 10.417e^{-0.2t} \qquad [2]$$

The coefficients in Eq. [2] differ from those in Eq. [1], but again the dominant time constant dominates the transient response.

Example 2.6   Determine the response to a unit step change of input of a system component with transfer function $5/(s+4)(s^2+2s+5)$.

Fig. 2.10 Response of $G(s) = 1/(1+s)(1+2s)(1+5s)$ to: (a) unit step input; (b) unit ramp input (Example 2.5)

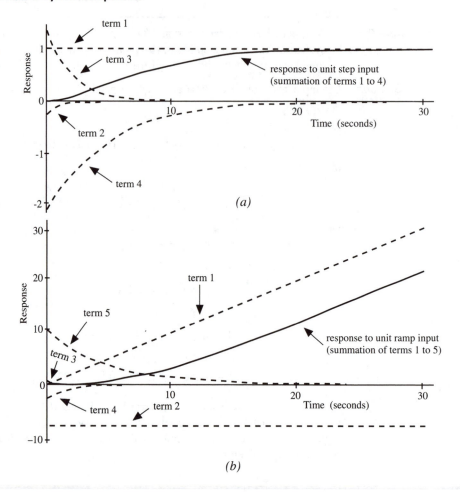

(a)

(b)

Solution   The Laplace transform of the output is

$$\theta_o(s) = \frac{5}{s(s+4)(s^2+2s+5)} = \frac{A}{s} + \frac{B}{s+4} + \frac{Cs+D}{s^2+2s+5}$$

$$= \frac{0.25}{s} - \frac{0.096}{s+4} - \frac{0.154s+0.692}{s^2+2s+5}$$

$$= \frac{0.25}{s} - \frac{0.096}{s+4} - \frac{0.154(s+1)}{(s+1)^2+2^2} - \frac{0.269(2)}{(s+1)^2+2^2}$$

$$\therefore\ \theta_o(t) = 0.25 - 0.096e^{-4t} - 0.154e^{-t}\cos 2t - 0.269e^{-t}\sin 2t$$

or

$$\theta_o(t) = 0.25 - 0.096e^{-4t} - 0.310e^{-t}\sin(2t+0.52)$$

The damped oscillation is of greater magnitude and decays more slowly than the exponential decay, as would be expected from the root positions of $-1 \pm j2$ and -4.

Note that whereas the transient response of a first order system can be described by reference to the value of the time constant, and that of a second order system by the values of ω_n and ζ, for a higher order system it is described by characteristics such as the following relating to a step input:

(i) rise time (time to rise from, say, 0.05 to 0.95 of the step size)
(ii) settling time (to within $\pm 2\%$, say)
(iii) maximum overshoot (as a percentage of the step size)
(iv) number of oscillations
(v) magnitude of steady state error

Consider now the frequency response of a higher order system. If both denominator polynomial $P_d(s)$ and numerator polynomial $P_n(s)$ are factorised then it will be seen that $G(s)$ can be factored into the product and quotient of just four forms of term, namely K, s, $(1 + \tau s)$ and $(s^2 + 2\zeta\omega_n s + \omega_n^2)/\omega_n^2$. $G(s)$ can thus be thought of as a gain together with a number of integral or derivative terms, of first order lags or leads, and of second order lags or leads. $G(j\omega)$ will factorise into a gain K and three forms of term $j\omega$, $1 + j\omega\tau$, $(\omega_n^2 - \omega^2 + j2\zeta\omega\omega_n)/\omega_n^2$ on numerator or denominator. The overall magnitude and phase can be obtained by multiplying magnitudes and adding phases of the constituent terms. Consider now what the magnitude and phase characteristics for these terms $G_n(j\omega)$ are:

(a) For a gain K which appears in the numerator as a multiplying factor the magnitude in db is $20\log_{10} K$, a constant and thus independent of ω, and there is no phase shift, hence the phase contribution is zero.
(b) A factor s on the denominator of $G(s)$ represents an **integral term**, and $G_i(j\omega) = 1/j\omega = -j/\omega$. This has magnitude $1/\omega$, which in decibels is $-20\log_{10}\omega$ and on the Bode plot is a straight line of slope -20 db per decade passing through 0 db at $\omega = 1$ r/s. $-j/\omega$ is imaginary and negative, thus the phase shift is $-90°$, a constant lag independent of ω. A factor s on the numerator represents **differentiation**. For this $G_n(j\omega) = j\omega$, hence the magnitude is $20\log_{10}\omega$, a line of slope $+20$ db per decade passing through 0 db at $\omega = 1$ r/s, and the phase shift is constant at $+90°$.
(c) The magnitude and phase for a first order lag, a term $(1 + j\omega\tau)$ on the denominator of $G_i(j\omega)$, were shown in Section 2.4 to be $-20\log_{10}\sqrt{1 + \omega^2\tau^2}$ and $-\tan^{-1}\omega\tau$ respectively, and to appear on the Bode plot as shown in Fig. 2.8(b). For a **first order lead**, a term $(1 + j\omega\tau)$ on the numerator of $G(j\omega)$, the magnitude is $+20\log_{10}\sqrt{1 + \omega^2\tau^2}$ and the phase is $+\tan^{-1}\omega\tau$. The Bode plots, Fig. 2.11(a), are mirror images about the 0 db and 0 degree axes of the plots for the lag.
(d) The magnitude and phase for a second order lag, a term $(\omega_n^2 - \omega^2 + j2\zeta\omega\omega_n)/\omega_n^2$ on the denominator of $G(j\omega)$, were shown in Section 2.3 to be $-40\log_{10}\sqrt{(1 - \omega^2/\omega_n^2)^2 + (2\zeta\omega/\omega_n)^2}$ and $-\tan^{-1}(2\zeta\omega/\omega_n)/(1 - \omega^2/\omega_n^2)$ respectively and to appear on the Bode plot as shown in Fig. 2.5(b). For a **second order lead**, a term of this form on the numerator, the magnitude and phase are numerically the same but of opposite sign. The Bode plots, Fig. 2.11(b), are therefore also mirror images about the 0 db and $0°$ axes of the plots for the equivalent lag.

A number of approaches can be used to find magnitude and phase for $G(j\omega)$:

(1) rationalise $G(j\omega)$ and find the real and imaginary components, or magnitude and phase, as functions of frequency. Insert frequency values and plot the results. This is tedious and prone to error.
(2) as (1) for each factor separately, then combine and plot on a polar plot. Some obvious errors can be detected. Less laborious.

Fig. 2.11 Bode plots: (*a*) simple lead; (*b*) quadratic lead

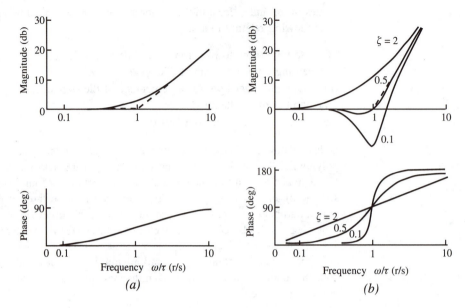

(3) draw components on a bode plot and add – as accurately as desired with the aid of standard curves.

(4) use a computer program – easiest, if available, but methods (2) and (3) help to fix understanding.

Example 2.7   A system has transfer function $G(s) = 100/(1 + s)(1 + 5s)(s^2 + s + 25)$. Sketch the general form of the polar diagram which can be expected from inspection of the transfer function. Establish expressions which give the cartesian and polar coordinates of points on the polar plot as functions of frequency. Use these to calculate the frequency response and to display it on a polar plot.

Solution

$$G(s) = \frac{100}{(1 + s)(1 + 5s)(s^2 + s + 25)}$$

At very low frequencies the lag will be zero, and the gain 4; at high frequencies the lag will be $90 + 90 + 180 = 360°$, and the gain will tend to zero. The second order component has $\omega_n = 5$ and $\zeta = 0.1$, hence a local increase in amplitude is expected around $\omega = 5$. This is at a higher frequency than the two first order components which have break points at $\omega = 1$ and $\omega = 1/5 = 0.2$, i.e. these will dominate. The general shape expected is as in Fig. 2.12(a).

$$G(j\omega) = \frac{100}{(1 + j\omega)(1 + 5j\omega)(25 - \omega^2 + j\omega)} = \dots$$

$$= \frac{100}{(25 - 132\omega^2 + 5\omega^4) + j(151\omega - 11\omega^3)}$$

$$\therefore \; G(j\omega) = \frac{100(25 - 132\omega^2 + 5\omega^4) - j100(151\omega - 11\omega^3)}{(25 - 132\omega^2 + 5\omega^4)^2 + (151\omega - 11\omega^3)^2}$$

Fig. 2.12 Polar plot, $G(s) =$

$$\dfrac{100}{\dfrac{(1+s)(1+5s)}{(s^2+s+25)}},$$

Examples 2.7 and 2.8: (*a*) general form; (*b*) true curve

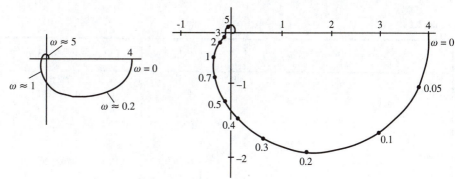

This expression gives the real and imaginary coordinate of the polar plot. The magnitude, $\sqrt{\text{real}^2 + \text{imag}^2}$, is

$$\frac{100}{\sqrt{(25 - 132\omega^2 + 5\omega^4)^2 + (151\omega - 11\omega^3)^2}}$$

and the phase, $\tan^{-1}\dfrac{\text{imag}}{\text{real}}$, is

$$-\tan^{-1}\left(\frac{151\omega - 11\omega^3}{25 - 132\omega^2 + 5\omega^4}\right)$$

For

$\omega = 0 \qquad G(j\omega) = 4 + j0$

$\omega = 0.2 \quad G(j\omega) = 1.522 - j2.324 \qquad \text{Mag} = 2.778 \quad \text{Phase} = -57°$

$\omega = 1 \qquad G(j\omega) = -0.340 - j0.467 \quad \text{Mag} = 0.578 \quad \text{Phase} = -126°$

$\omega = 3 \qquad G(j\omega) = -0.127 - j0.026 \quad \text{Mag} = 0.129 \quad \text{Phase} = -168°$

$\omega = 5 \qquad G(j\omega) = -0.037 + j0.152 \quad \text{Mag} = 0.157 \quad \text{Phase} = -256°$

etc.

The plot is shown in Fig. 2.12(b).

Example 2.8

Obtain the polar plot for the system transfer function of Example 2.7 by combining the frequency responses of the individual components to obtain the overall response.

Solution

The system can be thought of as four components in series: two unity gain simple lags $1/(1+s)$ and $1/(1+5s)$, a unity gain complex lag $25/(s^2+s+25)$, and a gain of 4. The magnitudes and phases for these can be obtained from relatively simple expressions, and the overall magnitude and phase can then be obtained by multiplication and addition respectively. This is best done in a tabular form, as in Table 2.1. Simple checks should be made, e.g. for the simple lags when $\omega = 1/\tau$ the magnitude should be 0.707 and the phase $-45°$, and for the complex lag when $\omega = \omega_n$ the magnitude should be $1/2\zeta$ and the phase $-90°$. Table 2.1 also highlights the relative dominance of the components.

The plot is that shown in Fig. 2.12(b).

Table 2.1

ω	0	0.2	1	3	5	10
$\lvert 4 \rvert$	4	4	4	4	4	4
$\left\lvert \dfrac{1}{1+j\omega} \right\rvert = \dfrac{1}{\sqrt{1+\omega^2}}$	1	0.981	0.707	0.316	0.196	0.100
$\left\lvert \dfrac{1}{1+j5\omega} \right\rvert = \dfrac{1}{\sqrt{1+25\omega^2}}$	1	0.707	0.196	0.067	0.040	0.020
$\left\lvert \dfrac{25}{(25-\omega^2)+j\omega} \right\rvert = \dfrac{25}{\sqrt{(25-\omega^2)^2+\omega^2}}$	1	1.002	1.041	1.536	5.000	0.330
Overall magnitude = product	4	2.780	0.577	0.130	0.157	0.003
$\angle \dfrac{1}{1+j\omega} = -\tan^{-1}\omega$	0	-11.3	-45.0	-71.6	-78.7	-84.3
$\angle \dfrac{1}{1+j5\omega} = -\tan^{-1}5\omega$	0	-45.0	-78.7	-86.2	-87.7	-88.9
$\angle \dfrac{25}{(25-\omega^2)+j\omega} = -\tan^{-1}\dfrac{\omega}{25-\omega^2}$	0	-0.5	-2.4	-10.6	-90.0	-172.4
Overall phase = sum	0	-56.8	-126.1	-168.4	-256.4	-345.6

Example 2.9   Draw a Bode plot for the system of Examples 2.7 and 2.8.

Solution   There are four components of the transfer function which will be plotted separately on the Bode plot, using straight line approximations to assist:

(i) gain term of 4: 12.04 db on magnitude plot, 0° on phase plot
(ii) simple lag $1/(1+s)$: break point at $\omega = 1$ r/s; magnitude approximation is 0 db to $\omega = 1$, then slope -20 db/decade; phase approximation is 0° to $\omega = 0.1$, straight line through $-45°$ at $\omega = 1$ to $-90°$ at $\omega = 10$, then constant $-90°$
(iii) simple lag $1/(1+5s)$: as for (ii) but centred on break point at $\omega = 1/5 = 0.2$ r/s
(iv) complex lag $25/(s^2+s+25)$: break point at $\omega = \omega_n = 5$ r/s; magnitude approximation is 0° to $\omega = 5$, then slope -40 db/decade; true magnitude as Fig. 2.5(b) for $\zeta = 0.1$; phase 0° for $\omega \ll 5$, $-90°$ when $\omega = 5$, $-180°$ for $\omega \gg 5$; true phase as Fig. 2.5(b) for $\zeta = 0.1$

The magnitude and phase curves for the four components are shown in Fig. 2.13. These are then added graphically to give the overall frequency response shown in the same figure.

Example 2.10   What is the harmonic response of a system with transfer function

$$G(s) = \frac{2(1+s)}{s(1+0.2s)(1+0.1s)}$$

Fig. 2.13 Bode plot,
$G(s) =$
$$\dfrac{100}{\dfrac{(1+s)(1+5s)}{(s^2+s+25)}},$$

Example 2.9

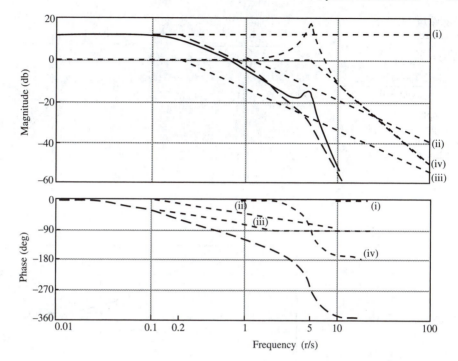

Solution   At very low frequencies the simple lags and the simple lead have no effect on phase or magnitude of the input signal, hence the output is $2/j\omega = -j(2/\omega)$, i.e. as $\omega \to 0$ phase $\to -90°$ and magnitude $\to \infty$. At very high frequencies the magnitude tends to zero and the phase tends to $-90\ -90\ -90\ +90 = -180°$. The lead with $\tau = 1$ is more dominant than the lags with $\tau = 0.2$ and 0.1, hence the overall phase lag will be less than $90°$ at frequencies around 1 r/s. The general shape of the polar plot in the absence of the phase lead would be of the form of the dotted line in Fig. 2.14(a), and with the phase lead would be of the form of the dashed line.

Frequency response information will be calculated using the tabular method, as shown in Table 2.2. With these results, supplemented by additional frequency values if a more accurate curve is required, the polar plot, Fig. 2.14(b), can be drawn. If the magnitude values are converted to decibels the Bode plot, Fig. 2.14(c), can be drawn directly. Alternatively the general form could have been established by use of straight line approximations, a more accurate curve drawn by making corrections for the errors near the corner frequencies, and if greater accuracy were needed locally, this could be obtained by calculation as above.

2.6   System identification

The earlier sections of this chapter have described how, for a system of known transfer function, the response to a given forcing function can be evaluated. Where a system component is available for experimental testing the converse process can be employed in order to confirm the form of a theoretically derived transfer function, and to determine or confirm numerical values for the parameters, or alternatively, for a so-called 'black box' system where the governing equations are not known, to derive a representative transfer function. The process is one of subjecting the components to, say, a

Fig. 2.14   Harmonic response,

$$G(s) =$$

$$\frac{2(1+s)}{s(1+0.2s)(1+0.1s)},$$

Example 2.10:
(a) approx. polar plot;
(b) actual polar plot;
(c) Bode plot

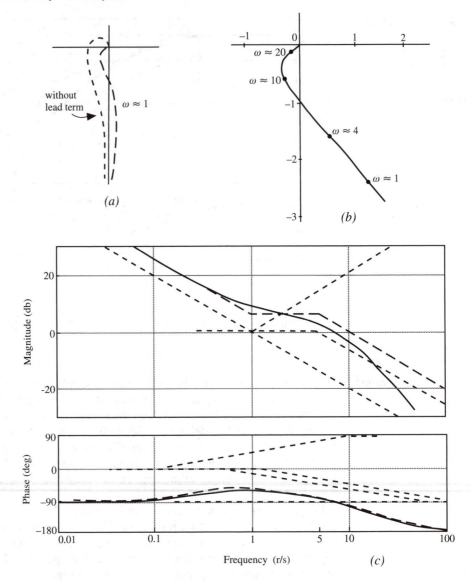

step change of input, recording the resultant output response, and by a process of curve fitting determining a transfer function which would yield this response. Alternatively, or additionally, the component may be tested with a sinusoidal input for a range of frequencies, the results plotted on polar and Bode plots, and curve-fitting undertaken.

If the response of a component to a step change of input has the general form of Fig. 2.7(a), i.e. it rises apparently exponentially to the new steady state and there is a sharp initial rise with the maximum slope at the time when the step is applied, then this suggests that a first order transfer function should be a good representation. The value of the time constant can be estimated by noting the time taken to reach 63% of the steady state value, and/or noting for tangents at various points the time to the point of intersection of the tangent with the steady state level (characteristics (ii), (iv) and (v) of Section 2.4). The steady state gain is the ratio of the steady state change in output to the input step size. If the frequency response for the same component is obtained it should

Table 2.2

ω	0	0.2	1	4	10	20
$\left\|\dfrac{2}{j\omega}\right\| = \dfrac{2}{\omega}$	∞	10	2	0.5	0.2	0.1
$\left\|\dfrac{1}{1+0.2j\omega}\right\| = \dfrac{1}{\sqrt{1+(0.2\omega)^2}}$	1	0.999	0.981	0.781	0.447	0.243
$\left\|\dfrac{1}{1+0.1j\omega}\right\| = \dfrac{1}{\sqrt{1+(0.1\omega)^2}}$	1	1.000	0.995	0.928	0.707	0.447
$\|1+j\omega\| = \sqrt{1+\omega^2}$	1	1.020	1.414	4.123	10.050	20.025
$|G(s)| = $ product	∞	10.190	2.760	1.494	0.635	0.218
$\angle \dfrac{2}{j\omega} = -90°$	-90	-90.0	-90.0	-90.0	-90.0	-90.0
$\angle \dfrac{1}{1+0.2j\omega} = -\tan^{-1}0.2\omega$	0	-2.3	-11.3	-38.7	-63.4	-76.0
$\angle \dfrac{1}{1+0.1j\omega} = -\tan^{-1}0.1\omega$	0	-1.1	-5.7	-21.8	-45.0	-63.4
$\angle 1+j\omega = +\tan^{-1}\omega$	0	11.3	45.0	76.0	84.3	87.1
$\angle G(j\omega) = $ sum	-90	-82.1	-62.0	-74.5	-114.1	-142.3

be close to a semicircle on a polar plot, Fig. 2.8(a), and should exhibit the characteristic shapes of Fig. 2.8(b) on the Bode plot. The value of τ can be estimated by determining the frequency for which the lag is 45°, τ being the reciprocal of this value. The gain can be estimated by determining the ratio of the output amplitude to the input amplitude for low values of frequency.

If the response of a component to a step change of input exhibits any overshoot, with or without subsequent oscillation, or if the maximum slope of the output response occurs later than the time of application of the step change, then the transfer function must be of higher order than first. Attempts should be made to see whether a second order transfer function would give a close fit to the response curve obtained experimentally. If the response is oscillatory then comparison with Fig. 2.4(a) will enable values of damping ratio ζ and undamped natural frequency ω_n to be estimated – the ratio of the peak output value to the steady state output change, and the time of occurrence of the peak, being good comparative indicators. The steady state gain is again the ratio of the steady state change in output to the input step size. The values of ζ, ω_n and gain can also be obtained from frequency response information by comparison of the polar or Bode plot with Fig. 2.5(a) or (b). Provided the fit is good then the system component can be closely represented by this second order transfer function. When the correspondence is less close then this procedure gives an estimate of the dominant complex roots. If the response is seen to be overdamped this implies that there will be two or more real roots, and if damping is large then study of the slope of the later part of the response will yield an estimate of the dominant time constant. An equivalent estimate can be obtained from the lower frequency region of the polar or

Bode plot. Evaluation of a second root is less easy, but it can be undertaken with the assistance of simulation or curve fitting computer software. If damping is close to critical then estimates of root values are rather approximate. For third and higher order systems additional roots are not easy to estimate.

Response testing of a practical system component can give a useful indication about linearity. The component can be considered linear provided the shape of the response curve resulting from a step change of input is not affected significantly by variation of the step size. The existence of a finite time delay (also referred to as a transport lag), which is not a non-linearity but which does require modification to standard linear analytical procedures, should also become evident from inspection of the step response.

2.7 More complex forcing functions

A linear system has been defined as one for which the principle of superposition applies, and application of this principle permits analytical determination of the response of a linear system to forcing functions which are rather less simple in form than the step, ramp, impulse and sinusoidal inputs considered earlier in this chapter. Typical forms of input for which analytical time responses can easily be derived are functions such as a pulse, and a ramp with a limiting value, which are shown in Fig. 2.15(a) together with three other functions of a similar nature. In all cases the input function can

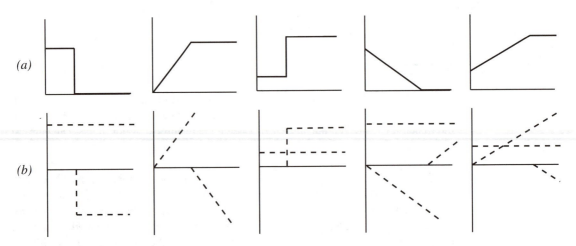

Fig. 2.15 Forcing functions which can be derived by superposition: (a) the forcing functions; (b) the step and ramp functions which are superposed

be created by superposition of simple steps and ramps with appropriate amplitudes, some of them with time delays, as shown in Fig. 2.15(b). A pulse of amplitude A and duration W, for example, results from the superposition of a step of amplitude A applied at time $t = 0$, and a second step of amplitude $-A$ applied at time $t = W$. The response to such a complex input function is obtained by summing the responses to the constituent simple functions in the manner illustrated by the next example.

Example 2.11   A system component has transfer function $2/(s + 1)(s + 0.2)$. Derive an expression for the response of the component to an input signal which increases suddenly by 3 units from a steady datum value, and then 6 seconds later falls suddenly back to the datum value.

Solution   It is helpful at the outset to consider qualitatively the response expected from inspection of the data given. The transfer function has two real roots and thus is second order and overdamped. Its significance is clearer when it is written in the form $10/(1+s)(1+5s)$, i.e. the steady state gain is 10 and the dominant time constant is 5 seconds. At $t=0$ a step of amplitude 3 is applied, and if there were no further change in input the output would rise to a steady value of $3 \times 10 = 30$ units, reaching within a few percent of this in a period slightly larger than $4 \times 5 = 20$ seconds (Fig. 2.16). At $t=6$ the input is removed. For a second order component the response cannot suddenly start to decrease; it will continue to a maximum shortly after $t=6$ and then fall back to zero with a settling time of a little more than 20 seconds, i.e. it will approach zero approximately exponentially becoming virtually zero at around $t=27$. This is shown qualitatively in Fig. 2.16.

Fig. 2.16   Qualitative response,

$G(s) = $
$\dfrac{10}{(1+s)(1+5s)}$,

Example 2.11

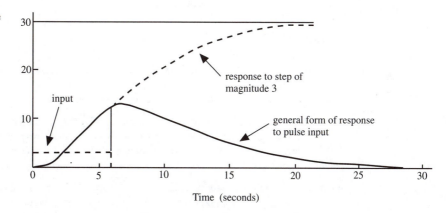

Time (seconds)

$$C(s) = \frac{2U(s)}{(s+1)(s+0.2)}$$

Consider first the period from $t=0$ to $t=6$
At $t=0$ a step of magnitude 3 is applied $\therefore$ $U(s) = 3/s$

$$\therefore \quad C(s) = \frac{6}{s(s+1)(s+0.2)} = \frac{A}{s} + \frac{B}{s+1} + \frac{D}{s+0.2}$$

$$A = 6/0.2 = 30$$
$$B = 6/(-1)(-0.8) = 7.5$$
$$D = 6/(-0.2)(0.8) = -37.5$$

$$\therefore \quad C(s) = \frac{30}{s} + \frac{7.5}{s+1} - \frac{37.5}{s+0.2}$$

$$\therefore \quad c(t) = 30 + 7.5e^{-t} - 37.5e^{-0.2t} \qquad \text{for } 0 \le t \le 6$$

$t(0) = 0$ and $t(6) = 18.72$

At $t=6$ the step is removed. Analytically this is effected by applying at this time a second step of magnitude -3, and superposing the responses. The response to this second step is

$$c(t') = -(30 + 7.5e^{-t'} - 37.5e^{-0.2t'}) \qquad \text{where } t' = t - 6$$

Superposition gives a response for the period from $t = 6$ onwards as

$$c(t) = (30 + 7.5e^{-t} - 37.5e^{-0.2t}) - (30 + 7.5e^{-(t-6)} - 37.5e^{-0.2(t-6)})$$

$$= 7.5e^{-t}(1 - e^6) - 37.5e^{-0.2t}(1 - e^{1.2})$$

$$\therefore\ c(t) = -3018e^{-t} + 87e^{-0.2t} \qquad \text{for } t \geq 6$$

$t(6) = 18.72$ and $t(\infty) = 0$

To find the peak value differentiate this expression and equate to zero.

$$\dot{c}(t) = 3018e^{-t} - 17.4e^{-0.2t}$$

$\dot{c}(t) = 0$ when $3018e^{0.2t} = 17.4e^t$, i.e. $e^{0.8t} = 173.4$, i.e. $t = 6.44$. The peak value is $c(6.44) = -3018e^{-6.44} + 87e^{-1.29} = 19.18$.

An accurate plot of output against time can be drawn from the two equations for $c(t)$.

Problems

1 For a component with transfer function $G(s) = 1/(1 + 0.1s)$ derive expressions for the output as functions of time for an input which is (a) a unit step change, (b) a unit impulse. Explain what is meant by the term 'settling time' and determine its value for the case of unit step change.

2 A system component is considered to be represented by the first order transfer function $G(s) = 1/(1 + 2s)$. Determine the response to an input which is increasing at (a) 1 unit per second, (b) 5 units per second. What is the magnitude of the steady state error in each case? How would the answers differ if the steady state gain were increased from 1 to 4?

3 For a system component with transfer function $G(s) = 6/(s^2 + 8s + 12)$ determine the response to (a) a unit step change of input, (b) an input step change of magnitude 10, (c) a unit ramp change of input. What is the magnitude of the steady state error in each case?

4 Write down the transfer function of a unity gain second order system with $\omega_n = 5$ and $\zeta = 0.4$. Using the Laplace transform approach find an expression for the response of this system to an input step change of unit amplitude. What is the maximum overshoot which occurs?

5 A system comprises three simple lags in series with gains of 1, 6, 2 and time constants 0.5, 1, 5 seconds respectively. Evaluate the output response $c(t)$ when, at time $t = 0$ with the component in a steady state condition, the input is increased by 10 units and held constant at that value. Evaluate $c(t)$ also for the case where the lags are in the reverse order.

6 For a third order component with transfer function $G(s) = 10/(1 + \tau s)(s^2 + s + 4)$ determine the response to a step input of unit magnitude when τ has a value of 1 second and 10 seconds respectively. Plot the curves, and comment on the results.

7 A system component has transfer function $G(s) = 5/(s^2 + 5s + 10)$. Without detailed calculation sketch first the general form expected for the frequency response when plotted as magnitude and phase against frequency. Calculate, subsequently, the values for magnitude and phase for the frequencies 0.2, 0.5, 1, 2, 5, 10 and 20 r/s.

8 By inspection sketch the general form of the polar plot for the transfer function $1/(s^3 + 5s^2 + 6s + 1)$, then calculate the harmonic response information for enough frequency values to permit the plot to be drawn more accurately.

9 Sketch the general form of the polar plot for a system with $G(s) = 1/(1 + 0.1s)(1 + 2s)(1 + 10s)$, and then draw it more accurately with the aid of some calculated points. Could the system be represented by a simpler transfer function? If so, what would it be?

10   Draw a Bode plot for the transfer function of Problem 9 using straight line approximations. Sketch in the true curve by estimating the errors in the regions of the corner frequencies. For a few sample frequencies check that the harmonic response characteristics obtained are the same as those derived in Problem 9.

11   By calculating harmonic response information for about eight appropriately spaced frequency values for each value of τ draw the polar plots for the system of Problem 6.

12   Draw the Bode plots for the system of Problem 6.

13   Evaluate the overall transfer function for the unity feedback system of Fig. P2.1, and hence find the response $c(t)$ to a unit step change of input $r(t)$. Plot this response. Draw also the polar diagram for the closed loop response $C(j\omega)/R(j\omega)$.

Fig. P2.1

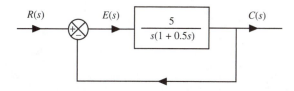

14   Plot in the s plane the poles of the transfer function $2/(1+0.5s)(1+0.2s)(s^2+0.4s+1)$. What does this suggest about the nature of the dynamic behaviour to be expected? Without calculating it analytically, sketch the general form which the unit step response can be expected to have, explaining the main characteristics.

15   Evaluate the response of a first order system with transfer function $G(s) = 2/(1+5s)$ to an input which rises suddenly by 1 unit, and then 4 seconds later by a further 2 units.

16   Determine the response as a function of time of a system with transfer function $1/(1+s)(1+5s)$ to (a) an input which rises at one unit per second until it reaches a value of 10 and then remains constant at this value, (b) a ramp input rising at 1 unit per second which after 10 seconds changes to a ramp rising at 3 units per second.

17   A system component has transfer function $G(s) = 1/(s^2+5s+4)$. Sketch the general shape of the response which you would expect if, starting with the component at rest, an input were applied in the form of a square wave with maximum value unity, minimum value zero, and frequency 0.1 Hz, and explain the main characteristics of the expected response.
    Evaluate the response analytically for the first 15 seconds after application of the input.
    Sketch the general shape expected if the frequency were to be the higher value of 0.5 Hz, giving reasoned argument for the shape drawn.

CHAPTER 3

ANALYSIS OF FEEDBACK SYSTEMS

Closed loop or feedback systems are utilised as regulators, to maintain the output variable close to some desired value, or as servomechanisms, to have the output following closely some desired time function. When the system is subjected to an input or disturbance function then the output will respond in a certain manner. Closing a loop around a process tends to result in a response which is more oscillatory than that of the process alone, and perhaps even unstable. For any given system, therefore, it is necessary to check that it is not unstable, and often also to obtain a measure of the margin from instability. Within the two approaches for studying system behaviour introduced in the last chapter, time response and frequency response analysis, there are techniques for assessing stability in both the time and frequency domains, and these are described in the first two sections of this chapter. Section 3.3 then shows how closed loop harmonic response information can be obtained from the open loop response, and explains the significance of the resultant performance characteristics. With a unity feedback system the error signal is the difference between the actual output and the input or desired output, and although transient errors are unavoidable it is generally desirable for the error in the steady state to be small. Section 3.4 describes how to calculate the steady state error and shows how it depends both on the input signal and on the system transfer function. Measures taken to improve the stability margin result in a deterioration in the steady state error, and vice versa, and hence design, to be considered in Chapter 4, is concerned with selecting a form of system which will meet the general objectives, and then achieving an acceptable compromise between stability and accuracy requirements.

3.1 Stability analysis – time domain

It was shown in the last chapter that the nature of the response of a system to a transient disturbance is primarily dependent on the values of the roots of the characteristic equation. In particular, if any root has a positive real part and thus is located in the right half of the complex s-plane then there is a component of the response which increases without bound and hence the system is unstable. For a closed loop system (Fig. 1.16, p. 23) the overall transfer function is

$$\frac{C(s)}{R(s)} = \frac{G(s)}{1 + G(s)H(s)} \qquad [3.1]$$

and the characteristic equation, obtained by equating the denominator to zero, is

$$1 + G(s)H(s) = 0 \qquad [3.2]$$

Analysis to determine whether or not the system is stable is a process either of determining the values of the roots and noting whether any has a positive real part, or of using a numerical technique to determine this without evaluating the roots.

For a first or second order characteristic equation the values of the roots can be determined directly by solution of the equation. For a third order characteristic equation one root must be real, and the other two may either both be real or be a complex conjugate pair. One real root can always be found by a trial and error iterative process, this factor then being divided out to leave a quadratic whose roots can be solved for directly. A fourth order characteristic equation may have four real roots, two real and one complex conjugate pair, or two complex conjugate pairs. Provided there are any real roots the same procedure could be used. Most engineers will have access to a polynomial solving computer program and this provides the easiest way of finding the roots of higher order equations. In all cases once the values of the roots have been found it can then be determined by inspection whether any roots have positive real parts, and hence whether the system is unstable. Furthermore, if stable, plotting of the roots in the s-plane will give an idea of how close the system is to instability by looking at the position of the dominant roots relative to the imaginary axis.

Example 3.1 A unity feedback system has forward loop transfer function $5/s(s+1)(s+3)$. Is it stable?

Solution The characteristic equation is $1 + 5/s(s+1)(s+3) = 0$, i.e. $s(s^2 + 4s + 3) + 5 = 0$, i.e. $s^3 + 4s^2 + 3s + 5 = 0$.

$$
\begin{array}{llll}
\text{Try} \quad s = -1 & \text{LHS} = -1 + 4 - 3 + 5 & = 5 \\
s = -2 & \text{LHS} = -8 + 16 - 6 + 5 & = 7 \\
s = -3 & \text{LHS} = -27 + 36 - 9 + 5 & = 5 \\
s = -4 & \text{LHS} = -64 + 64 - 12 + 5 & = -7 \\
s = -3.5 & \text{LHS} = \ldots & = 0.625 \\
s = -3.55 & \text{LHS} = \ldots & = 0.021 \\
s = -3.56 & \text{LHS} = \ldots & = -0.100 \\
s = -3.552 & \text{LHS} = \ldots & = -0.0037 \\
\end{array}
$$

The summation for $s = -1, -2, -3$, and -4 can be done by mental arithmetic. It is clear that there should be a value between -3 and -4 which satisfies the equation, and with the aid of a calculator one can rapidly home in on $s = -3.552$. Working to two decimal places divide out a factor $(s + 3.55)$ from the cubic:

$$(s + 3.55)(s^2 + as + b) = 0$$

Now a will have a value 0.45 to give the correct coefficient of s^2, i.e. 4. Then

$3.55 \times 0.45 + b = 3$, the coefficient of s, i.e. $b = 1.40$. And as a check that there is no obvious error: $1.4 \times 3.55 = 4.9 \approx 5$

$$\therefore \quad (s + 3.55)(s^2 + 0.45s + 1.40) = 0$$

The roots of the quadratic are

$$1/2(-0.45 \pm \sqrt{((0.45)^2 - 4(1)(1.4))}),$$

i.e. $-0.22 \pm j1.16$. These roots are in the left half plane but close to the imaginary axis (Fig. 3.1).

$$\phi = \tan^{-1}(1.16/0.22) = 79.3° \quad \text{and} \quad \zeta = \cos\phi = 0.186$$

Fig. 3.1 Root positions, Example 3.1

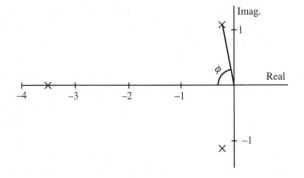

The system is stable but, dominated by the complex roots, is relatively oscillatory, and very close to second order with $\omega_n \approx 1.16$, $\zeta \approx 0.2$. Had all three roots been real the trial and error process might have homed in on one of the other roots.

Routh–Hurwitz criterion

If no polynomial solving routine is available then a numerical technique referred to as the **Routh–Hurwitz criterion** can be used to determine with a modest amount of effort whether or not a system is stable, without the need to calculate the values of the roots. The method, which will be described without proof, is a combination of two similar techniques developed independently by Routh and Hurwitz.

Step 1: Write down the characteristic equation as a polynomial in s in the form

$$a_0 s^n + a_1 s^{n-1} + a_2 s^{n-2} + \cdots + a_{n-1} s + a_n = 0 \qquad [3.3]$$

Step 2: Examine this equation with its descending powers of s and check that all coefficients are positive and that no coefficient is zero – a necessary but not sufficient condition for stability. If any term is negative or missing then there is at least one root with positive real part, the system is unstable, and it is not necessary to proceed to the next step.

Step 3: Arrange the coefficients in two rows labelled s^n and s^{n-1} thus:

s^n	a_0	a_2	a_4	a_6	$\cdot\ \cdot$
s^{n-1}	a_1	a_3	a_5	$\cdot\ \cdot$	

Step 4: Using these coefficients calculate numerical values for $n - 1$ additional rows to form a tabular array which tapers to a single value for the s^0 row and has the form:

s^n	a_0	a_2	a_4	a_6
s^{n-1}	a_1	a_3	a_5	
s^{n-2}	b_1	b_2	b_3	
s^{n-3}	c_1	c_2		
$\vdots$	$\vdots$			
s^0	f_1			

Each of these values is calculated using four numbers from the two rows immediately above – the pair one column to the right, and the pair in the first column, combined in the following way:

$$
\begin{array}{ccc}
a_0 & a_2 & \cdot \\
a_1 & a_3 & \cdot \\
b_1 & & \cdot
\end{array}
\qquad
\begin{array}{ccc}
a_0 & \cdot & a_4 \\
a_1 & \cdot & a_5 \\
& b_2 &
\end{array}
\qquad
\begin{array}{ccc}
a_0 & \cdot & a_6 \\
a_1 & \cdot & a_7 \\
& & b_3
\end{array}
$$

$$
b_1 = a_2 - \frac{a_0 a_3}{a_1} \qquad b_2 = a_4 - \frac{a_0 a_5}{a_1} \qquad b_3 = a_6 - \frac{a_0 a_7}{a_1}
$$

$$
\begin{array}{ccc}
a_1 & a_3 & \cdot \\
b_1 & b_2 & \cdot \\
c_1 & & \cdot
\end{array}
\qquad
\begin{array}{ccc}
a_1 & \cdot & a_5 \\
b_1 & \cdot & b_3 \\
& c_2 &
\end{array}
$$

$$
c_1 = a_3 - \frac{a_1 b_2}{b_1} \qquad c_2 = a_5 - \frac{a_1 b_3}{b_1} \qquad \text{etc.}
$$

Step 5: Examine the first column of values, i.e. $a_0, a_1, b_1, c_1, d_1, \ldots$ For the system to be stable all these values must be positive. Every change of sign indicates one root with positive real part.

Example 3.2   A feedback system has a process with transfer function $G(s) = 5/s(s+1)(s+3)$ in the forward loop, and transducer with transfer function $H(s) = 1/(s+1)$ in the feedback loop. Is it stable?

Solution   The characteristic equation is

$$
1 + \frac{5}{s(s+1)(s+3)} \frac{1}{(s+1)} = 0
$$

i.e.

$$
s(s+1)(s^2 + 4s + 3) + 5 = 0
$$

i.e.

$$
s^4 + 5s^3 + 7s^2 + 3s + 5 = 0
$$

All terms are present and positive, so proceed to form the Routh array and calculate coefficients:

s^4	1	7	5
s^3	5	3	
s^2	6.4	5	
s^1	-0.9		
s^0	5		

$$
b_1 = 7 - \frac{(1)(3)}{5}, \quad b_2 = 5 - \frac{(1)(0)}{5}, \quad b_3 = 0
$$

$$
c_1 = 3 - \frac{(5)(5)}{6.4}, \quad c_2 = 0 - \frac{(5)(0)}{6.4}
$$

$$
d_1 = 5 - \frac{(6.4)(0)}{-0.9}
$$

There are two changes in sign ($+$ to $-$ and $-$ to $+$) in the first column, hence the system is unstable. Note that the negative number is relatively small, suggesting the system may not be far from stability. If the forward loop gain of 5 were reduced to 4 then the value c_1 would be -0.125 and the system would still be unstable, but if reduced to 3.5 then c_1 would be 0.266 and the system would be stable, but very oscillatory.

A problem arises if a value of zero is calculated for the first number on one of the rows since the array cannot then be completed owing to the need to divide by zero when trying to calculate the numbers for the following row. For the rare cases where the remaining numbers in the row with the zero are non-zero the problem can be overcome by a change of variable, by replacing s by $1/r$ in the characteristic equation; this yields an equation with the coefficients reversed, to which the normal procedure can then be applied. More commonly, there is a complete row of zeros, and this usually indicates a condition of marginal stability which is discussed next.

Critical value of gain for marginal stability

As loop gain increases a system of order higher than second will become more and more oscillatory, and beyond a certain value of gain will be unstable. The Routh–Hurwitz criterion is particularly useful for finding this critical value, and for calculating the frequency of oscillation at this limiting condition. The procedure is to write down the characteristic equation with the gain included as a variable K, to complete the Routh array in which some terms will be expressions involving K, to note which of the first column terms are positive for small values of K and negative for large values of K, and for each to calculate the critical value of K for which the sign changes. For high order systems more than one value may result, and the critical value will be the lower. As will be seen in Example 3.4 there are systems where there is a minimum critical value of gain in addition to the maximum, and these are referred to as **conditionally stable systems**. When the gain K is set to the critical value, then a complete row of zeros appears in the Routh array. The frequency of oscillation can be found by solving the **auxiliary equation** which is defined as the equation obtained from the coefficients taken from the row above the row of zeros, with terms in s as indicated by the power of s at the left hand end, which will be even. These procedures are illustrated by two numerical examples.

Example 3.3   What is the limiting value of process gain beyond which the system of Example 3.2 becomes unstable? What is the frequency of oscillation at that value?

Solution   The characteristic equation is

$$1 + \frac{K}{s(s+1)(s+3)} \frac{1}{(s+1)} = 0$$

i.e.

$$s^4 + 5s^3 + 7s^2 + 3s + K = 0$$

The Routh array is

$$
\begin{array}{lccc}
s^4 & 1 & 7 & K \\
s^3 & 5 & 3 & \\
s^2 & 6.4 & K & \\
s^1 & 3 - 5K/6.4 & & \\
s^0 & K & &
\end{array}
$$

If K is negative then there is one change of sign showing that the system would be unstable, and this would have been clear from the negative term in the characteristic equation. For small positive values of K the term $3 - 5K/6.4$ is positive, and for large values of K it is negative, the limiting value for stability being given by

$$
3 - \frac{5K_{\text{crit}}}{6.4} = 0, \text{ i.e. } K_{\text{crit}} = 3.84
$$

When K has this value then the s^1 row contains only zeros. The auxiliary equation, formed from the coefficients in the s^2 row above is

$$
6.4s^2 + 3.84 = 0
$$

$$
\therefore \ s = \pm \text{j}0.775
$$

Hence the system is stable for all values of K in the range $0 < K < 3.84$, and when $K = 3.84$ the system will oscillate without any damping at a frequency of 0.775 r/s, i.e. 0.123 Hz.

Example 3.4   For what values of gain K, if any, is the feedback system shown in Fig. 3.2 stable?

Fig. 3.2 Block diagram, Example 3.4

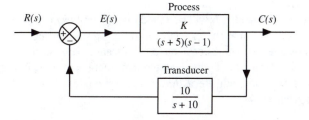

Solution   By inspection it can be seen that the process in the forward loop is second order, which, owing to the minus sign in the denominator and in its characteristic equation $s^2 + 4s - 5 = 0$, is unstable. Also, the transducer is first order, with time constant 0.1 s and gain unity. The characteristic equation for the closed loop system is

$$
1 + \frac{K}{(s+5)(s-1)} \frac{10}{(s+10)} = 0
$$

i.e.

$$
s^3 + 14s^2 + 35s + (K - 50) = 0
$$

By inspection it can be seen that if $K < 50$ the closed loop is unstable. The Routh array is

$$
\begin{array}{lll}
s^3 & 1 & 35 \\
s^2 & 14 & K - 50 \\
s^1 & 35 - (K - 50)/14 & 0 \\
s^0 & K - 50 &
\end{array}
$$

For stability $35 - (K - 50)/14 > 0$, i.e. $540 - K > 0$, i.e. $K < 540$ and also $K - 50 > 0$, i.e. $K > 50$. Hence the feedback system is stable for $50 < K < 540$, and is thus conditionally stable. When $K = 540$ the frequency of oscillation is given by the auxiliary equation $14s^2 + 490 = 0$ i.e. $s = \pm\sqrt{35}$, i.e. 5.9 r/s. When $K = 50$ there is a positive real root, and the system output tends to infinity without oscillation.

This example illustrates that an open loop unstable system can be made stable by putting a feedback loop round it. Note, however, that if the loop is opened accidentally or due to malfunction then runaway will occur, and the output will increase until either some component fails or the amplitude is limited by saturation.

Critical values for marginal stability of other parameters

The procedure can also be used to find critical values of parameters other than loop gain, e.g. in Example 3.2 with transducer transfer function $H(s) = 1/(1 + Ts)$ for a gain of 5 the characteristic equation is

$$
Ts^4 + (4T + 1)s^3 + (3T + 4)s^2 + 3s + 5 = 0
$$

When, as happens here, the variable appears in several coefficients, particularly those of high order terms, then some terms in the lower rows of the Routh array are rather involved and great care is necessary to avoid errors of arithmetic. Creation of the Routh–Hurwitz array and evaluation of the critical value(s) of T will be left as an exercise for the reader. It will be found that the first term in the s^2 row would be zero for $T = -\frac{1}{3}$ or -1, and the first term in the s^3 row would be zero for $T = -0.318$ or $\frac{1}{2}$. Only positive real values of T have significance here, so the system will be stable for $0 \leq T < 0.5$. (If a pair of complex numbers emerges from equating a first column term to zero this will not indicate a limiting condition for stability for the parameter being investigated.)

3.2 Stability analysis – frequency domain

The frequency response of a very oscillatory system has the general characteristic shown in Fig. 3.3. As the forcing frequency increases a region of resonance is approached where there is considerable increase in output amplitude and a corresponding rapid increase in phase lag, and beyond the frequency at which the amplitude attains a maximum there is rapid attenuation and a flattening of the phase curve. The higher the peak amplitude the closer the system is to instability, with the amplitude tending to infinity at the frequency found by the Routh–Hurwitz approach for the condition of marginal stability.

Fig. 3.3 Frequency response of a very oscillatory system

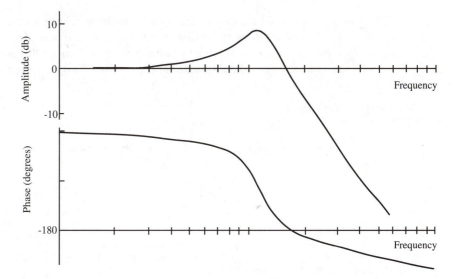

Frequency domain analysis techniques for feedback systems involve study of the open loop frequency response, the response $G(j\omega)H(j\omega)$ when the loop is opened at a point such as shown in Fig. 3.4, and the prediction from this of the closed loop

Fig. 3.4 Feedback system with loop opened

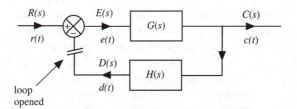

behaviour. If the open loop system is forced with a unit amplitude sine wave as the input $r(t)$ then, provided the system is linear, the signal $d(t)$ at the break will also be a sine wave, one which is lagging the input and has an amplitude which is normally not unity. Provided the open loop system $G(s)H(s)$ is of order higher than second then there is a certain frequency for which the phase lag of $d(t)$ is 180°, and if the loop were closed this sine wave when inverted would be in phase with the input and would reinforce it. If the amplitude of $d(t)$ at this frequency is greater than unity, then the signal as it passes around the loop will build up without limit, and the loop will be unstable. If the amplitude of $d(t)$ is less than unity then the signal will build up to a constant amplitude. Study, therefore, of the open loop amplitude for the frequency at which the open loop phase is $-180°$ tells whether the closed loop system is stable or not, and can give a quantitative measure of how close the system is to instability.

Nyquist stability criterion

On the polar plot of the open loop harmonic response of a system, the negative real axis represents a phase lag of 180° and the $(-1, j0)$ point on it represents a lag of 180° with an amplitude of unity, and thus is a critical point as regards stability in the way that has been described in the paragraph above. For a feedback system where the open loop is stable, a simplified form of the **Nyquist stability criterion** states that the closed loop will also be stable provided that the open loop polar plot does not enclose the $(-1, j0)$ point, that it will be marginally stable if it passes through this point, and unstable if it encloses it. Figure 3.5 illustrates this by showing the open loop polar plots for a typical

Fig. 3.5 Illustration of Nyquist plot

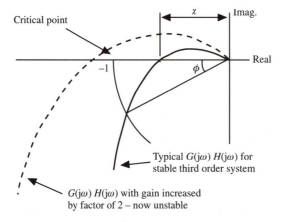

third order system for two values of gain. For the lower value of gain the harmonic response locus does not enclose the critical point, the amplitude for 180° lag is around 0.6, and the system would be stable with the loop closed. If the gain is increased by a factor of 2, then the locus is scaled up without change of shape and now encloses the critical point, and the system would be unstable with the loop closed. For a certain value of gain between these the locus would pass through the critical point and the system would be marginally stable. Such a polar plot for an open loop system, where the $(-1, j0)$ point has particular significance, is referred to as a **Nyquist plot**. If the open loop system is unstable then this simplified form of the Nyquist criterion is not valid, and the more complex full form of the criterion, which is described in many comprehensive texts on control engineering, must be applied.

Gain margin and phase margin

It should be clear from the above that for a stable system the closer that the Nyquist plot approaches the critical point, the nearer the system is to instability. Two measures of the degree of stability are given by the gain margin and phase margin. The **gain margin** is the amount by which the gain can be increased before the system becomes unstable. It has the value $1/x$, where x is the magnitude of $G(j\omega)H(j\omega)$ corresponding to a phase of $-180°$ (Fig. 3.5). It is normally quoted in decibels (db), i.e. $GM = 20 \log_{10}(1/x)$ db. The **phase margin** is the amount of additional lag which can be introduced before the system becomes unstable, i.e. the angle by which the phase of $G(j\omega)H(j\omega)$ is short of $-180°$ when the magnitude is unity; $PM = $ angle ϕ (Fig. 3.5).

If the open loop harmonic response information is presented on a Bode plot rather than a polar plot, then gain and phase margins can be read off as shown in Fig. 3.6. To determine the phase margin find the frequency at which the magnitude is 0 db (i.e. unity), read off the phase and note by how many degrees it is above the $-180°$ level (i.e. by how much the lag is short of 180°). To determine the gain margin, find the frequency for which the phase is $-180°$, read off the magnitude for this frequency, and

Fig. 3.6 Measurement of gain margin and phase margin from Bode plot

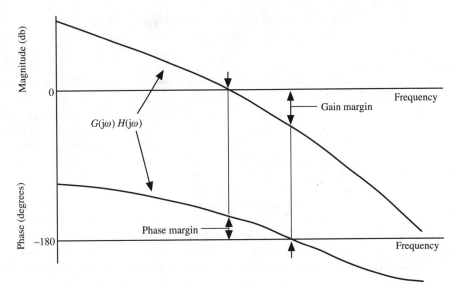

note how many db this is below zero, i.e. by how much the gain can be increased before the magnitude is unity, 0 db. Increase of loop gain K causes the magnitude plot to be raised vertically with no change to the phase plot, which can be seen to decrease both the phase margin and the gain margin.

Example 3.5   By means of the Nyquist stability criterion, determine whether the unity feedback system of Example 3.1, with forward loop transfer function $5/s(s+1)(s+3)$, is stable and, if so, determine the values of the gain margin and phase margin.

Solution   The general shape to be expected for the Nyquist plot, from inspection of $G(s)H(s) = 5/s(s+1)(s+3)$, is that shown in Fig. 3.7(a). The plot need only be drawn for a range

Fig. 3.7 Nyquist plot, Example 3.5

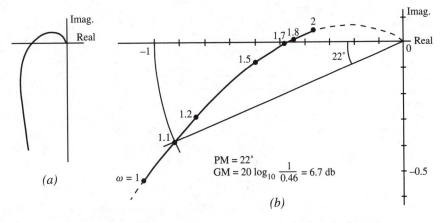

$$PM = 22°$$
$$GM = 20 \log_{10} \frac{1}{0.46} = 6.7 \text{ db}$$

of frequency values which encompasses both the $-180°$ phase and the unity magnitude conditions. Rather than rationalise $G(j\omega)H(j\omega)$ to find the real and imaginary parts in terms of ω, the overall magnitude and phase is best obtained by multiplying the magnitude and adding the phase values for the constituent factors as in Table 3.1. The frequency range of interest, with an integration term and two time

Table 3.1

Frequency ω (r/s)	1	2	1.1	1.2	1.5	1.7	1.8
$\lvert 5/3j\omega \rvert = 5/3\omega$	1.667	0.833	1.515	1.389	1.111	0.980	0.926
$\lvert 1/(1+j\omega) \rvert = 1/\sqrt{1+\omega^2}$	0.707	0.447	0.673	0.640	0.555	0.507	0.486
$\lvert 1/(1+0.33j\omega) \rvert = 1/\sqrt{1+(\omega/9)^2}$	0.994	0.976	0.993	0.991	0.986	0.983	0.981
$\lvert G(j\omega)H(j\omega) \rvert = \text{product}$	1.171	0.363	1.012	0.881	0.608	0.488	0.441
$\underline{/5/3j\omega} = -90°$	-90.0	-90.0	-90.0	-90.0	-90.0	-90.0	-90.0
$\underline{/1/(1+j\omega)} = -\tan^{-1}\omega$	-45.0	-63.5	-47.7	-50.2	-56.3	-59.5	-61.0
$\underline{/1/(1+0.33j\omega)} = -\tan^{-1}0.33\omega$	-18.4	-33.7	-20.1	-21.8	-26.5	-29.5	-31.0
$\underline{/G(j\omega)H(j\omega)} = \text{sum (degrees)}$	-153.4	-187.2	-157.8	-162.0	-172.8	-179.0	-182.0

constants of 1 and 0.33 seconds, can be anticipated as being between about 1 r/s where the phase will be $(-90° - 45° - <45°)$, i.e. $< -180°$ and 3 r/s with phase $(-90° - >45° - 45°)$, i.e. $> -180°$. Having calculated values for 1 and 2 r/s, columns 1 and 2 of Table 3.1, it is clear that both unity magnitude and $-180°$ phase lie in this range, and appropriate intermediate frequency values can easily be decided upon. From the resultant plot, Fig. 3.7(b), since the harmonic locus does not enclose the critical -1 point, the system is seen to be stable with phase margin measured as $22°$ and gain margin $20\log_{10}(1/0.46) = 6.7$ db. The gain margin is reasonable but the phase margin is small and the system would be relatively oscillatory.

Generally it can be assumed that if the phase margin is around $40°$ then there will be a reasonable degree of stability, and the transient response would be acceptable. This is equivalent to a gain margin of around 6 db where the polar plot has the general form of Fig. 3.5 in the vicinity of the critical $(-1, j0)$ point. If not, as in Fig. 3.8(a), then a

Fig. 3.8 Systems with different forms of Nyquist plot: (a) where gain margin and phase margin values used by themselves can give false impression about stability; (b) conditionally stable system

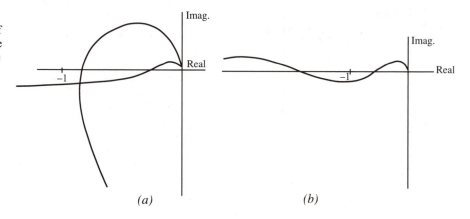

(a)   (b)

single value of either gain margin or phase margin can give a misleading indication of system stability, and both values must be given, e.g. a system can have a high gain margin and yet be close to instability. A combination of good phase margin and poor gain margin would represent an unusual system. The conditionally stable system, first encountered in Section 3.1, would have an open loop harmonic locus which crossed the negative real axis twice, as in Fig. 3.8(b). As shown the locus does not enclose the critical $(-1, j0)$ point and the closed loop will therefore be stable, but if the gain is either reduced by about 50% or more, or increased by about 100% or more then the critical point will be enclosed and the closed loop will be unstable.

Example 3.6   A unity feedback system has forward loop transfer function $K/s(s^2+s+4)$. Determine the values of gain margin and phase margin when: (a) $K=1$; (b) $K=3$.

Solution   For a value of $K=1$ either calculate frequency response information and plot on a polar or Bode plot, or draw the Bode plot with the aid of straight line approximations and standard second order response curves, as in Section 2.5. The resultant curve is shown in Fig. 3.9. From this $GM = 20 \log_{10} 1/0.25 = 12.0$ db, and $PM = 86$ degrees. For

Fig. 3.9 Nyquist plot, Example 3.6

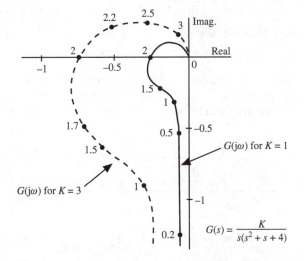

$K=3$ the locus on the polar plot is scaled up by a factor of 3, or the magnitude curve on the Bode plot is shifted up vertically by $20 \log_{10} 3 = 9.54$ db. With this higher gain, $GM = 20 \log_{10} 1/0.75 = 2.5$ db, and $PM = 77$ degrees. The gain margin is now relatively small; transient behaviour would be relatively oscillatory. Slight increase of gain to $K=3.5$ would give a rapidly reducing value of phase margin.

3.3 Closed loop frequency response

For a single block the harmonic response or frequency response, described in Sections 2.3 to 2.5, gives the magnitude and phase of the steady state output signal relative to a sinusoidal input forcing function. The block could represent a single component with some transfer function $G(s)$, but alternatively could represent a closed loop system (or even a multi-loop system) which has been reduced to a single block by the methods of Section 1.5. Section 3.2 considered the harmonic response of an 'open loop system', a feedback system where the loop has been opened and which is equivalent to a single block with transfer function $G(s)H(s)$. The information which was obtained from this, whether the system is stable or not together with values of gain and phase margins, is however information about the response of the system with the loop closed. The open loop frequency response is thus used to assess the relative stability of the closed loop system.

The closed loop frequency response, that of a block with transfer function $G(s)/(1 + G(s)H(s))$, helps to give understanding of dynamic system behaviour and is particularly relevant for amplifiers and servomechanisms. Closed loop frequency response can be determined from a knowledge of the open loop response, as will be shown now. For a unity feedback system:

$$\frac{C(s)}{R(s)} = \frac{G(s)}{1 + G(s)} \quad \therefore \quad \frac{C(j\omega)}{R(j\omega)} = \frac{G(j\omega)}{1 + G(j\omega)}$$

Now the overall closed loop magnitude, conventionally represented by the symbol M and often referred to as the **magnification**, is

$$M = \frac{|C(j\omega)|}{|R(j\omega)|} = \left| \frac{C(j\omega)}{R(j\omega)} \right| = \left| \frac{G(j\omega)}{1 + G(j\omega)} \right| = \frac{|G(j\omega)|}{|1 + G(j\omega)|} \tag{3.4}$$

and the overall closed loop phase is

$$\phi = \underline{/C(j\omega)} - \underline{/R(j\omega)} = \underline{/C(j\omega)/R(j\omega)} = \underline{/G(j\omega)} - \underline{/1 + G(j\omega)} \tag{3.5}$$

The closed loop magnitude is thus the ratio of the lengths of the vectors $G(j\omega)$ and $1 + G(j\omega)$, and the closed loop phase is the difference in the angles of these vectors. Figure 3.10 shows these vectors for two values of frequency ω for the system of

Fig. 3.10 Open loop frequency response of unity feedback system, showing vectors $G(j\omega)$ and $1 + G(j\omega)$

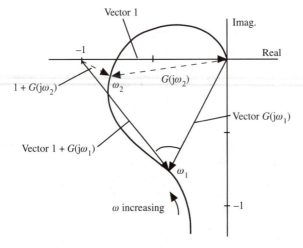

Example 3.6 with $K = 3$. It can be seen that $1 + G(j\omega)$ is the sum of a unit vector and $G(j\omega)$. At low ω $|G(j\omega)| \approx |1 + G(j\omega)|$ $\therefore$ $M \approx 1$, and $\underline{/G(j\omega)} \approx \underline{/(1 + G(j\omega))}$ $\therefore$ $\phi \approx 0$. As frequency increases $|1 + G(j\omega)|$ becomes smaller relative to $|G(\omega)|$ and at some frequency ω_p the magnification M will reach a peak value M_p in the region nearest to the $(-1, j0)$ point, with $\phi \approx 180°$. As $\omega \to \infty$ then $M \to 0$ and $\phi \to 270°$. M and ϕ can be evaluated for a range of frequencies by measurement from the polar plot, and the results displayed on a polar plot, Fig. 3.11(a), or on separate plots of M and ϕ against ω, Fig. 3.11(b).

It should be evident that many points on the complex plane of a polar diagram would

Fig. 3.11 Closed
loop frequency
response derived from
Fig. 3.9 for $K=3$: (*a*)
polar plot; (*b*) Bode
plot (linear scales)

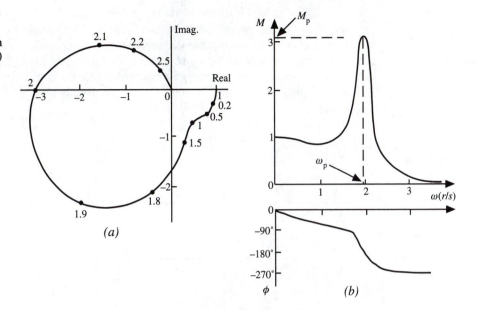

yield the same value of M, and that these could be joined to form a locus of constant M
value. Such loci, referred to as M contours, are circles of radius $M/(1 - M^2)$ with centre
at $(M^2/(1 - M^2), j0)$, as shown in Fig. 3.12. Loci of constant phase, referred to as N

Fig. 3.12 M and N
contours

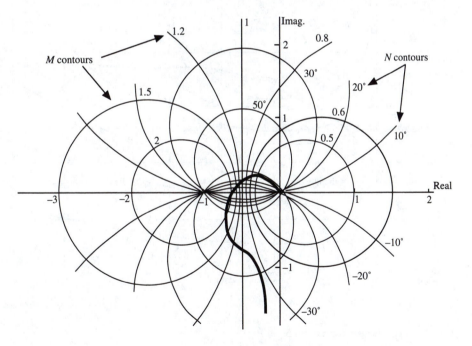

contours where $N = \tan \phi$, are also circles, of radius $0.5\sqrt{(1 + 1/N^2)}$ with centre at
$(-0.5, j0.5/N)$. If the M and N contours are shown on a polar diagram on which $G(j\omega)$
has been plotted then the closed loop information can be read off directly by noting the
intersections of $G(j\omega)$ with the contours. The reader should confirm that doing this on
Fig. 3.12 would yield points on the plots of Fig. 3.11.

A **Nichols chart** is an alternative method of presentation from which both open loop and closed loop harmonic information can be read. The rectangular axes represent open loop magnitude (db) and phase (degrees), and superimposed are the closed loop M and N contours. With such a diagram change of system gain involves simply shifting the $G(j\omega)$ locus upwards (increase) or downwards (decrease). Performance characteristics such as phase margin, gain margin, M_p, ω_p and bandwidth (see below) can be read off directly, Fig. 3.13.

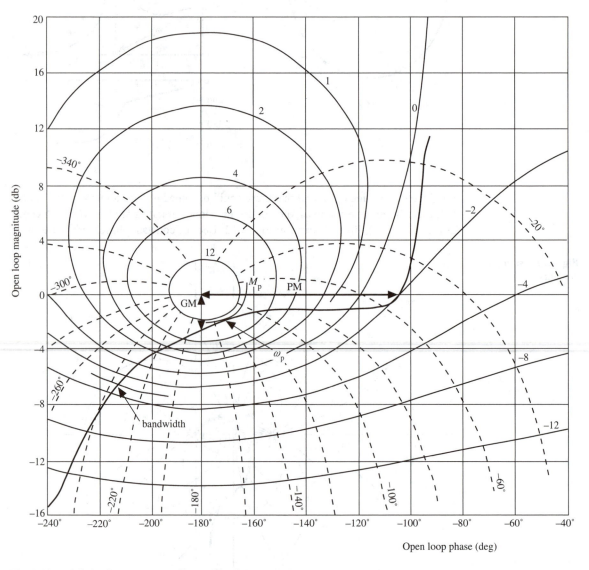

Fig. 3.13   Nichols chart, corresponding to Fig. 3.11.

For a feedback system such as an electronic amplifier for boosting a signal voltage, or a servomechanism such as that to drive the stylus on a pen recorder, the ideal closed loop response, when plotted as in Fig. 3.11(b) would be to have $M = $ constant and $\phi = 0$ for the range of signal frequencies of interest, and $M = 0$ for higher noise frequencies.

This ideal is not attainable. The following **closed loop harmonic response performance characteristics** indicate the nature of the actual response:

gain
peak magnification M_p
ω_p
bandwidth – the frequency at which $M = -3$ db (0.707).

If the system is overdamped then there is no peak and M_p and ω_p are not defined. A value of M_p in the range 1.1–1.5 is generally considered to be acceptable. Provided the input signal frequency is half ω_p or less then the response is relatively flat. Any unwanted signal (noise) at frequencies in excess of the bandwidth will be attenuated and thus largely filtered out by the system. The steeper the slope of the M curve at the bandwidth frequency the sharper will be the cut-off characteristics.

Much of the above relates to a unity feedback system. If $H(s) \neq 1$ then some changes are needed.

$$M = \frac{|G(j\omega)|}{|1 + G(j\omega)H(j\omega)|} \text{ and } \phi = \underline{/G(j\omega)} - \underline{/1 + G(j\omega)H(j\omega)}$$

Hence to obtain the closed loop response from a polar plot the locus $G(j\omega)H(j\omega)$ must also be drawn, and the vector $G(j\omega)$ compared with the vector $1 + G(j\omega)H(j\omega)$.

Example 3.7   For the unity feedback system of Example 3.1 find the values of M_p, ω_p and bandwidth. By how much should the gain be altered so that $M_p = 1.3$? What would then be the bandwidth?

Solution   Information must first be calculated to enable the open loop harmonic response to be plotted. The tabular approach will be used rather than the less accurate graphical approach via a Bode plot. Table 3.2 gives enough values to enable $G(j\omega)$ to be plotted and to show what intermediate frequency values would be useful. The most convenient plot is on a Nichols chart, Fig. 3.14. From this, by interpolation, it can be estimated that $M_p = 8$ db $= 2.5$, $\omega_p = 1.2$ r/s, and bandwidth $= 1.7$ r/s. For a value of $M_p = 1.3$ shift the $G(j\omega)$ locus vertically until it is just tangential to the M contour for $M = 1.3 = 2.28$ db. This requires a shift downward by 6.7 db, i.e. a decrease of gain by a factor of 2.16 from 5 to 2.3. The new bandwidth is 1.2 r/s.

3.4 Steady state error

Closed loop systems, it has been seen, employ feedback in order to ensure that the output of a process is sensibly constant (regulator action) or to control the process so that the output is maintained close to some desired varying time function (servo-mechanism action). A signal representing the actual output is fed back and subtracted from the input to the closed loop, which represents the desired output, to yield an error signal which is then used to determine the direction and amount of corrective action to be applied at the process input (Figs. 1.1(b), 1.2, 4.8). Ideally the error should be zero at all times, but this can never be achieved. Following any change of input, transiently there will be an error which cannot be avoided, although there is a measure of control over the magnitude and duration of the transient error through ensuring an appropriate

Table 3.2

Frequency ω (r/s)	0.1	0.2	0.3	0.5	0.7	1	1.5	2	3
$\lvert 5/j\omega \rvert = 5/\omega$	50	25	16.67	10.00	7.14	5.00	3.33	2.50	1.67
$\lvert 1/(1+j\omega) \rvert = 1/\sqrt{1+\omega^2}$	0.995	0.981	0.958	0.894	0.819	0.707	0.555	0.447	0.316
$\lvert 1/(3+j\omega) \rvert = 1/\sqrt{9+\omega^2}$	0.333	0.333	0.332	0.329	0.325	0.316	0.298	0.277	0.229
$\lvert G(j\omega) \rvert = 20\log_{10}$ (product)db	24.38	18.24	14.49	9.37	5.58	0.96	-5.18	-10.19	-18.35
$\angle 5/j\omega = -90°$	-90.0	-90.0	-90.0	-90.0	-90.0	-90.0	-90.0	-90.0	-90.0
$\angle 1/(1+j\omega) = -\tan^{-1}\omega$	-5.7	-11.3	-16.7	-26.6	-35.0	-45.0	-56.3	-63.4	-71.6
$\angle 1/(3+j\omega) = -\tan^{-1}\omega/3$	-1.9	-3.8	-5.7	-9.5	-13.1	-18.4	-26.6	-33.7	-45.0
$\angle G(j\omega) = $ sum	-97.6	-105.1	-112.4	-126.1	-138.1	-153.4	-172.9	-187.1	-206.6

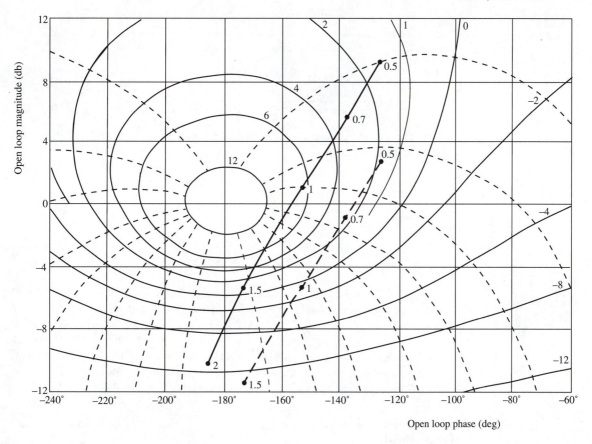

Fig. 3.14   Nichols chart, Example 3.7

degree of stability. Also, after the transient has died out, there may remain a steady state component of error and the system must be designed to ensure that this is kept to an acceptably low value. This section explains the nature of steady state error, how its magnitude can be evaluated, and how in general it can be reduced or eliminated.

Consider, first, a unity feedback system in which the signal fed back is the actual process output and hence the input signal is the desired output, Fig. 3.15.

$$E(s) = R(s) - C(s)$$

and $C(s) = G(s)E(s)$

$$\therefore \ E(s) = R(s) - G(s)E(s)$$

$$\therefore \ E(s) = \frac{R(s)}{1 + G(s)} \tag{3.6}$$

Now, applying the final value theorem of Laplace transform analysis, the steady state error is given by

$$\lim_{t \to \infty} e(t) = \lim_{s \to 0} sE(s) = \lim_{s \to 0} \frac{sR(s)}{1 + G(s)} \tag{3.7}$$

The steady state error depends, therefore, on the form and the magnitude both of the system transfer function $G(s)$ and of the system input function $R(s)$. The types of input

Fig. 3.15 Unity
feedback system

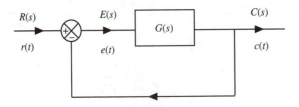

function of most interest are a steady input (the result of applying a step change, say)
and a ramp input.

Steady input

For a unit step change $R(s) = 1/s$

$$\therefore \ \lim_{t\to\infty} e(t) = \lim_{s\to 0} \frac{1}{1 + G(s)} = \frac{1}{1 + \lim_{s\to 0} G(s)} \qquad [3.8]$$

If $G(s) = K/(s+1)(s+5)$, say, then the steady state error for a unit input would be
$5/(K+5)$, i.e. the output would be $1 - 5/(K+5) = K/(K+5)$ rather than unity. It can be
seen that increase in the gain K will result in a reduction in the steady state error. If
$G(s) = K/s(s+1)(s+5)$, say, then the steady state error for a unit input would be
$1/(1 + \infty) = 0$, i.e. the output would also be unity, irrespective of the value of gain K. If
$G(s) = K/s^2(s+1)(s+5)$, say, then the steady state error is also zero. The value
$\lim_{s\to 0} G(s)$ is referred to as the **positional error constant**, K_p.

Ramp input

For a unit ramp change $R(s) = 1/s^2$

$$\therefore \ \lim_{t\to\infty} e(t) = \lim_{s\to 0} \left(\frac{1}{s + sG(s)} \right) = \frac{1}{\lim_{s\to 0} sG(s)} \qquad [3.9]$$

For $G(s) = K/(s+1)(s+5)$ the steady state error would be $1/0 = \infty$. For $G(s) = K/s(s+1)(s+5)$ the steady error would be $5/K$, a finite error which would decrease
with increase in K. When steady state is reached the output will be increasing at the
same rate as the input, but lagging by $5/K$. For $G(s) = K/s^2(s+1)(s+5)$ the steady state
error would be zero. The value $\lim_{s\to\infty} sG(s)$ is referred to as the **velocity error
constant** or **velocity error** K_v, and has units second^{-1}.

Acceleration input

For an acceleration input $r(t) = \frac{1}{2}t^2$, $R(s) = 1/s^3$

$$\therefore \ \lim_{t\to\infty} e(t) = \lim_{s\to 0} \left(\frac{1}{s^2 + s^2 G(s)} \right) = \frac{1}{\lim_{s\to 0} s^2 G(s)} \qquad [3.10]$$

The steady state errors for the three forms of transfer function considered above are ∞,
∞ and $5/K$ respectively. The value $\lim_{s\to\infty} s^2 G(s)$ is the **acceleration error
constant**, or **acceleration error** K_a and has units second^{-2}.

One conclusion to be drawn from the above is that the form of the steady state error is determined by the number of factors s in the denominator of $G(s)$. If there is no factor s (referred to as a Type 0 system) then for a steady input there is a finite error, and for a ramp or otherwise changing input, there is an infinite error. If there is a single factor s (Type 1 system) then for a steady input the error is zero, for a ramp input it is finite, and for an acceleration input it is infinite. If there is a double factor s^2 (Type 2 system) then the error is zero for both a steady and a ramp input and finite for an acceleration input. In all cases the magnitude of the steady state error, if finite, decreases with increase in gain K. For inputs of magnitude other than unity, the magnitude of any finite error would simply be in proportion.

If the closed loop system does not have unity feedback, $H(s) \neq 1$, then the 'error' signal must be defined differently:

$$E(s) = R(s) - C(s)H(s)$$

and

$$C(s) = G(s)E(s)$$

$$\therefore \ E(s) = \frac{R(s)}{1 + G(s)H(s)}$$

The above conclusions remain valid, with $G(s)$ replaced by $G(s)H(s)$.

Example 3.8   For the feedback system of Example 3.1, with $G(s) = 5/s(s+1)(s+3)$, $H(s) = 1$ what is the steady state error for the following inputs $r(t)$: (i) 10; (ii) 0.5t; (iii) 5t^2; (iv) $1 + 0.5t$.

Solution   Steady state error is $\lim\limits_{t \to \infty} e(t) = \lim\limits_{s \to 0} \dfrac{sR(s)}{1 + G(s)}$

(i) $R(s) = 10/s \ \therefore \ e(\infty) = \lim\limits_{s \to 0} \dfrac{10}{1 + 5/s(s+1)(s+3)} = \dfrac{10}{\infty} = 0$

(ii) $R(s) = 0.5/s^2 \ \therefore \ e(\infty) = \lim\limits_{s \to 0} \dfrac{0.5}{s + 5/(s+1)(s+3)} = \dfrac{0.5}{5/3} = 0.3$

(iii) $R(s) = 10/s^3 \ \therefore \ e(\infty) = \lim\limits_{s \to 0} \dfrac{10}{s^2 + 10s/(s+1)(s+3)} = \dfrac{10}{0} = \infty$

(iv) $R(s) = \dfrac{1}{s} + \dfrac{0.5}{s^2} \ \therefore \ e(\infty) = 0.3$, directly or by superposition.

Problems

1   Find the values of the roots of each of the following equations, and hence suggest the form of system behaviour which can be expected if these are system characteristic equations:

(a) $s^3 + 3.7s^2 + 4s + 1.2 = 0$

(b) $s^3 + 4s^2 + 7s + 6 = 0$

(c) $s^3 + s^2 - s + 15 = 0$

(d) $s^4 + 6s^3 + 15s^2 + 22s + 12 = 0$

2 By using the Routh–Hurwitz technique determine for each of the characteristic equations of Problem 1 whether the system it represents is stable or unstable.

3 Has the equation $s^5 + s^4 + 4s^3 + 4s^2 + 6s + 3 = 0$ any roots with positive real parts?

4 A feedback system has forward loop transfer function $G(s)$ and a transfer function $H(s)$ in the feedback path which have the values $G(s) = K/s(1 + 2s)(1 + 10s)$, $H(s) = 0.1/(1 + 0.5s)$. For what values of gain K is the system stable? If there is a critical value of K what would be the frequency of oscillation at that value?

5 Repeat Problem 4, but with $G(s) = K/(s^2 + 5s + 8)(s^2 + 3s + 16)$ and $H(s) = 1/(s + 1)$.

6 Repeat Problem 4 but with $G(s) = K/(s + 15)(s^2 - s + 5)$ and $H(s) = (s + 1)/(s + 10)$.

7 A unity feedback system has forward loop transfer function $40/(s + 5)(s^2 + 2s + 4)$. Use the Routh–Hurwitz criterion to determine whether or not the loop is stable. Determine by calculation the harmonic response information for enough frequency values to enable a plot to be drawn. Draw a Nyquist plot and from it confirm the stability condition determined earlier. If the system is stable, what are the values of gain margin and phase margin?

8 The frequency response of a process obtained experimentally is given in the following table:

Frequency (r/s)	0.2	0.4	0.6	0.8	1	1.5	2	2.5
Magnitude	11.55	5.80	3.90	2.92	2.35	1.60	1.21	0.95
Phase lag (deg)	95	100	105	110	116	130	146	163

Frequency (r/s)	3	4	5	7	10
Magnitude	0.74	0.41	0.22	0.07	0.02
Phase lag (deg)	182	217	243	274	297

Plot the information on a polar diagram, and hence determine the values of the gain margin and phase margin which would characterise the response if the process were to be placed in a closed loop with unity gain negative feedback. Could a value of 50° be obtained for the phase margin by alteration of the process gain only? If so what change would be required for the gain, and what would be the corresponding gain margin? Could a value of 6 db be attained for the gain margin by alteration of the process gain only? If so, what change would be required for the gain, and what would be the corresponding phase margin?

9 Repeat Problem 8, using a Bode plot in place of the polar plot.

10 How would the results for Problems 8 and 9 differ if in the feedback there were included a transducer (Fig. P3.1) and the transducer had a first order transfer function with unity gain and time constant 0.2 second?

11 Starting from the open loop polar plot drawn for Problem 8 evaluate closed loop harmonic response values and draw the closed loop polar plot. Estimate from this the values of M_p and ω_p for the loop. Transfer the results to a Bode plot and use this to estimate the values of M_p and ω_p, and also the bandwidth.

Fig. P3.1

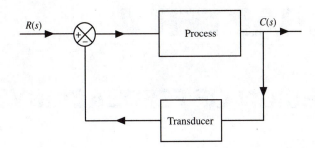

12 Obtain the values of M_p, ω_p and bandwidth for the closed loop system of Problem 8 by plotting the harmonic response information on a Nichols chart.

13 Using the frequency response information calculated for Problem 7 plot the harmonic response on a Nichols chart, and from it estimate the values of M_p, ω_p and bandwidth. By how much should the gain be altered to achieve a value of $M_p = 1.3$, and what would then be the bandwidth?

14 A unity feedback system has forward loop transfer function $G(s) = 25/s(s^2 + 5s + 16)$. What is the magnitude of the steady state error for: (a) a unit step change of input; (b) a unit ramp input; (c) a ramp input of 0.2 units per second?

15 A unity feedback system has a process with transfer function $G(s) = K/(1 + s)(1 + 6s)$ in the forward loop. What is the magnitude of the steady state error? How would this change if there were a phase compensation element $G_c(s) = (1 + 6s)/(1 + 0.8s)$ in series with the process (Fig. P3.2)?

Fig. P3.2

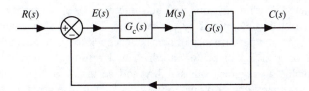

16 The forward loop of a unity feedback servomechanism comprises a motor with transfer function $3/s(s + 5)(s + 15)$ and amplifier with gain K. What value must K have to achieve a velocity error coefficient of 5 second^{-1}? What is then the magnitude of the steady state error?

CHAPTER 4

DESIGN OF FEEDBACK SYSTEMS

When a control system is to be designed, the process to be controlled will generally have a certain transfer function determined by analysis and/or testing, and there may be other existing elements such as transducers whose dynamic characteristics also cannot be altered. Within any loop will be some form of amplification or gain whose magnitude must be determined by the designer. The first step in the design process is to investigate the performance of a basic feedback loop, and determine whether a suitable value of gain K can be found which will result in dynamic characteristics that are acceptable as regards both stability and accuracy. Such a system, where corrective action is directly proportional to the error signal, is referred to as one with proportional control action. If the performance requirements cannot be met thus, then additional elements can be included to modify the loop behaviour in such a way that the response becomes acceptable. The chapter starts by considering root locus diagrams, which are a useful aid to understanding the effect on system behaviour of parameter variations, and of changes to the form of the system. Proportional control action is largely described in this section. Section 4.2 then describes the benefits to be gained by the inclusion of integral action and/or derivative action in a two term or three term controller. Also included is an explanation of rate feedback, which is a form of derivative action. Section 4.3 describes an alternative approach to improving system behaviour, that of introducing a phase compensation network. In a book of this size and one concentrating on essentials it is only possible to describe the most basic techniques of control engineering design and analysis, so the final section of this chapter is an introduction to some of the more advanced techniques.

4.1 Root locus diagrams

The **root locus diagram** is a useful aid to understanding for the system designer. It is a plot of the locus of the positions in the s plane of the roots of the characteristic equation as the gain K (or some other variable) varies from zero to infinity. It shows which roots are dominant, how close the other roots are, and how the positions of the dominant roots vary as K varies. The reader may have access to a computer program which when supplied with the values of the system coefficients or the poles and zeros, calculates and plots the root positions for a range of values of gain. Failing this a polynomial solving routine can be used to evaluate the roots, which can then be plotted manually. Before the advent of computers W R Evans devised a graphical method for creating root locus plots, referred to as the **root locus technique**, which employed a set of aids to construction to narrow down the area of search for points on the loci. The technique, which is now outlined, remains useful since, with a small amount of experience, the expected form of the plot can quickly be sketched freehand. The most useful aids to construction are listed without proof, and the few more laborious ones omitted, on the assumption that accurate plots will be computer generated.

The characteristic equation is $1 + G(s)H(s) = 0$ with $G(s)H(s)$ usually available in factorised form, hence the characteristic equation can be written as

$$\frac{K(s - z_1)(s - z_2)\ldots(s - z_m)}{(s - p_1)(s - p_2)(s - p_3)\ldots(s - p_n)} = -1 \qquad [4.1]$$

$z_1, z_2, \ldots, z_m$ are called **zeros** and $p_1, p_2, p_3, \ldots, p_n$ are called **poles**, values of s which would make the function $G(s)H(s)$ zero and infinite respectively. Any value of s which satisfies this equation will be a root of the characteristic equation and hence will lie on a locus. It should be appreciated that since s, the zeros, and the poles all represent points in the complex s plane then a term such as $(s - p_1)$ represents the vector from p_1 to s. Equation 4.1 can thus be written as two separate equations which must be satisfied, the **angle condition** and the **magnitude condition**:

$$(\underline{/s - z_1} + \underline{/s - z_2} + \ldots \underline{/s - z_m}) - (\underline{/s - p_1} + \underline{/s - p} + \ldots \underline{/s - p_n})$$
$$= \text{odd multiple of } 180° \qquad [4.3]$$

$$\frac{K|s - z_1|\ |s - z_2|\ldots|s - z_m|}{|s - p_1|\ |s - p_2|\ |s - p_3|\ldots|s - p_n|} = 1 \qquad [4.3]$$

The former determines whether any given point s lies on a locus, and if it does the latter determines the corresponding value of K. Trial and error application of the angle condition to successive locations in the s plane will find points on the root loci, but the **aids to construction** listed below markedly reduce the area in which a search need be made, and in the case of No. 6 accurately determine certain sections of locus. It is suggested that an approximate plot be always sketched when designing a control loop, and if desired specific areas of interest be drawn in more accurately by use of protractor and ruler on graph paper (or more easily on a draughting machine, if available).

For any physical system the order of the numerator of $G(s)H(s)$ cannot exceed the order of the denominator, i.e. $n \geq m$, and the aids to construction will be described on this assumption:

1. The number of loci, one for each root of the characteristic equation, is equal to the number of poles n; the loci are symmetrical about the real axis, since complex roots occur in conjugate pairs.
2. The loci start $(K = 0)$ at the n poles.
3. m loci finish $(K = \infty)$ at the zeros, while the remaining $n - m$ approach $n - m$ asymptotes and move to infinity.
4. The angles of the asymptotes are 180° for 1; 90° and 270° for 2; 60°, 180°, and 300° for 3; 45°, 135°, 225° and 315° for 4; 36°, 108°, 180°, 252° and 324° for 5; etc.

5. The asymptotes intersect at a point on the real axis given by

$$\frac{\text{(sum of poles)} - \text{(sum of zeros)}}{\text{number of asymptotes}}$$

6. Those portions of real axis which have an odd number of (poles + zeros) on the axis to the right are portions of locus.
7. When two loci approach each other along the real axis they will meet and then break away from the axis at $\pm 90°$ to it.
8. The angle of departure from a complex pole, or arrival at a complex zero, can quickly be found graphically by applying the angle condition to a point very close to the pole or zero.
9. The crossing of the imaginary axis can be found from the auxiliary equation during application of the Routh–Hurwitz criterion.

The procedure, the general nature of the resulting root locus plots, and interpretation of their significance is now illustrated by numerical examples.

Example 4.1   Produce a root locus plot for a unity feedback system with forward loop transfer function $G(s) = K/s(s+1)(s+3)$. (When $K = 5$ this is Example 3.1.)

Solution   There are three poles, at $s = 0$, $s = -1$, $s = -3$, and no zeros. Hence there will be three loci which start at these poles and which all end at infinity. The loci will become asymptotic to lines at $60°$, $180°$ and $300°$ which intersect on the real axis at $(0 - 1 - 3)/3 = -4/3$. On the s plane mark the position of each pole with an x and draw the asymptotes (Fig. 4.1).

There will be portions of loci on the real axis between 0 and -1, since there is one pole on the axis to the right, and from -3 to $-\infty$, since there are three poles to the right. Draw these in.

Two portions of locus approach each other from $s = 0$ and $s = -1$; these will meet, separate at $\pm 90°$ and follow mirror image paths towards the asymptotes as shown. The points at which these loci cross the imaginary axis can be found from the Routh–Hurwitz method.

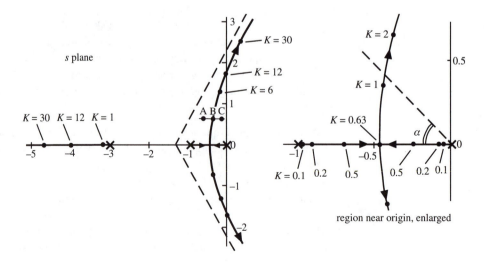

Fig. 4.1 Root locus plot for $G(s)H(s) = K/s(s+1)(s+3)$

The characteristic equation is $s^3 + 4s^2 + 3s + K = 0$ and the array is then:

$$
\begin{array}{ccc}
s^3 & 1 & 3 \\
s^2 & 4 & K \\
s^1 & 3 - K/4 & \\
s^0 & K &
\end{array}
$$

The critical value of K is given by $3 - K/4 = 0$, i.e. $K = 12$. The roots are then given by $4s^2 + 12 = 0$, i.e. $s = \pm j\sqrt{3}$. The loci thus cross the imaginary axis at $\pm j1.732$, when $K = 12$. Also $s^3 + 4s^2 + 3s + 12 = (s^2 + 3)(s + 4) = 0$, so the third root when $K = 12$ is at $s = -4$.

The general form of the root locus plot has been obtained very easily. A more exact plot in this case can be obtained by manual solution of the characteristic equation for a number of values of K, or from a polynomial solving program. Alternatively, by measurement to find the value of the left side of Eq. 4.2 for sets of points such as A, B and C, and interpolation to find the point at which the angle condition Eq. 4.2 is satisfied, a more exact path for the locus can be determined. It is helpful to graphically check the magnitude condition, Eq. 4.3, for the known critical points: the distances to the three poles can be measured as 1.73, 2.0, 3.45 and for $K/(1.73 \times 2 \times 3.45)$ to be equal to unity then K must be 11.94 which with more accurate measurement would be the value of 12. For the third root position $K/(1 \times 3 \times 4) = 1$, $\therefore K = 12$. Having checked correct application of the method, values of K can now be determined for a selection of points along the loci. A number of values of K are shown on Fig. 4.1.

Consider now what insight the root locus plot gives to the system behaviour: For very low values of K there are three roots, all on the negative real axis, and very close to the poles; there will be no overshoot and the root near the origin will dominate a response which is sluggish and similar to a first order response with largish time constant. Consider $K = 0.2$ for which the values of the roots are -0.074, -0.894 and -3.032. The overall transfer function is then

$$
\frac{C(s)}{R(s)} = \frac{G(s)}{1 + G(s)} = \frac{0.2}{s(s+1)(s+2) + 0.2} = \frac{0.2}{(s+0.074)(s+0.894)(s+3.032)}
$$

or

$$
\frac{C(s)}{R(s)} = \frac{1}{(1 + 13.5s)(1 + 1.12s)(1 + 0.33s)} \tag{4.4}
$$

The second root is 12 times as far from the origin as the dominant root, and thus the overall transfer function can be approximated by

$$
\frac{C(s)}{R(s)} = \frac{1}{1 + 13.5s} \tag{4.5}
$$

Shown in Fig. 4.2 are the initial portions of the step responses for the transfer functions of Eqs 4.4 and 4.5. The shapes of the curves differ only in the first two seconds or so, which is a short time in comparison to the settling time of about 50 seconds. As K increases the real roots near $s = 0$ and $s = -1$ approach one another, that near the origin becomes less dominant, and the response will become less sluggish and predominantly second order overdamped. When K increases beyond 0.63 the dominant roots become a complex conjugate pair and the predominantly second order behaviour will start to show tendency to overshoot. As K increases towards 12 the response becomes more and more oscillatory, and closer to a pure second order underdamped response as the third root to the left of $s = -3$ becomes less significant relative to the dominant roots. As K approaches 12 $\omega_n \rightarrow 1.73$ r/s, and for $K > 12$ the system is unstable. A selection of step response traces obtained by simulation is shown in Fig. 4.2.

Fig. 4.2 Unit step response traces for the system of Fig. 4.1

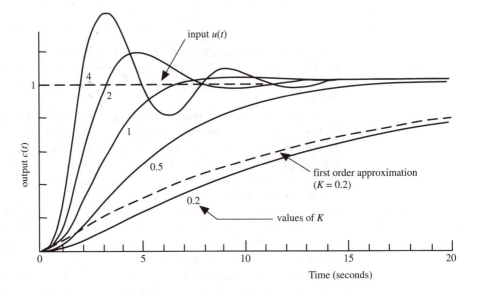

A damping factor of 0.7 for the dominant roots, $\alpha = 45°$ ($\zeta = \cos \alpha$, Section 2.5) in Fig. 4.1, requires $K \approx 1.2$. When $K = 1$ the step response has overshoot of 4% and settling time of about 13 seconds which is little different to that of a second order system, so the third root at $s = -3.3$, about 7 times farther from the origin than the dominant roots, has negligible effect. The 'best' response is likely to result for a value of K in the range 1 to 1.5, with overshoot increasing as rise time and settling time are reduced. The closed loop steady state gain is unity, but note that the forward loop gain is not K but $K/3$, since

$$G(s) = \frac{K}{s(s+1)(s+3)} = \frac{K/3}{s(1+s)(1+0.33s)}$$

$K = 1.2$ thus represents a forward loop gain of 0.4.

Example 4.2

Produce a root locus plot for a unity feedback system with forward loop transfer function $G(s) = K(s+1.5)/s(s+1)(s+3)$, i.e. Example 4.1 with the addition of a lead term.

Solution

As before there are three poles at $s = 0$, $s = -1$, and $s = -3$, but there is now also a zero (marked O) at $s = -1.5$. There are still three loci. One finishes at the zero, and the other two end at infinity, tending to asymptotes at 90° and 270° which intersect the real axis at $\{(-1-3)-(-1.5)\}/2 = -1.25$. The real axis from -3 to -1.5 will be a locus, and that between -1 and 0 will be two portions of locus which meet and break away at right angles moving to the asymptotes, Fig. 4.3(a).

There is unlikely to be any crossing of the imaginary axis, and this can be confirmed by the Routh–Hurwitz criterion. If a locus were to cross into the right half plane then it would need to recross to approach an asymptote.

Fig. 4.3 Root locus plot for:

(a) $G(s)H(s) = \dfrac{K(s+1.5)}{s(s+1)(s+3)}$

(b) $G(s)H(s) = \dfrac{K(s+0.5)}{s(s+1)(s+3)}$

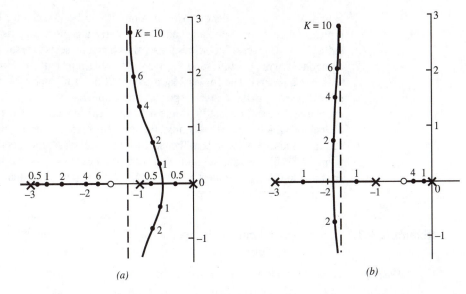

(a)

(b)

For very low values of K the root positions are little different to Example 4.1 and the response is sluggish (Fig. 4.4(a)). As K increases the response changes from predominantly first order to predominantly second order overdamped, to predominantly second order underdamped but with increasing influence from the third root. As K becomes very large the system remains stable with ω_n of the dominant roots becoming large and ζ small. Dominant roots with $\zeta = 0.7$ can be seen to require $K \approx 2$; ω_n at 1.1 r/s and forward loop gain $K/2 = 1$ are both about double the corresponding values in Example 4.1, which explains the faster response.

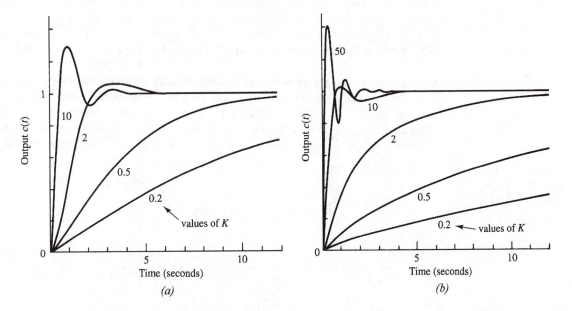

(a)

(b)

Fig. 4.4 Unit step response traces for the systems of Fig. 4.3

For a larger lead time constant, say 2 second, the root locus plot would be as shown in Fig. 4.3(b). As before the system is stable for all values of K, but now the response is likely always to be sluggish, dominated by the real root near the origin. For larger values of K some tendency to oscillation should be detectable. Figure 4.4(b) shows representative step responses, and these clearly are more sluggish than those in Fig. 4.4(a). $K=10$ gives a 'good' response, which can be seen to result from a second order component with $\zeta=0.57$ and $\omega_n=3.4$ r/s superimposed on a first order component with $\tau\approx2$ second. The forward loop gain is $K/6=1.67$, and this, with the larger value of ω_n, results in the faster response where the rise time is less than 1 second, the overshoot 3%, and the settling time 4.6 seconds. Reduction of K to 8 results in a peak overshoot which lies below unity. Increase of K to 15 results in a reduction of the settling time to 3.5 seconds at the expense of an increase in overshoot to 13%. When K is increased further, although the real root very near to $s=-0.5$ would be expected to continue to dominate it does not do so, since the amplitude of this first order component of response becomes progressively smaller.

Example 4.3   Sketch the form of the root locus plot for a system with $G(s)H(s)=K/(1+s)$ $(1+0.1s)(s^2+6s+18)$.

Solution $G(s)H(s)$ must first be written in the form

$$G(s)H(s) = 10K/(s+1)(s+10)(s^2+6s+18)$$

There are four poles at $s=-1$, $s=-10$, $s=-3\pm j3$, and no zeros. There are four loci, all of which end at infinity, tending to asymptotes at 45°, 135°, 225°, and 315° which intersect the real axis at $(-1-10-3-3)/4=-4.25$.

Two loci approach each other along the real axis from $s=-10$ and $s=-1$, then break away towards two of the asymptotes. The other two start at the complex poles and tend towards the remaining two asymptotes.

The angle of departure of the loci from the complex poles can be determined graphically: imagine a trial point s_1 a *very* small distance from $-3+j3$. Let $\underline{/s_1-(-3+j3)}=\theta$. By measurement $\underline{/s_1-(-3-j3)}\approx90°$, $\underline{/s_1-(-1)}\approx$ $\underline{/(-3+j3)-(-1)}\approx124°$, and $\underline{/s_1-(-10)}\approx\underline{/(-3+j3)-(-10)}\approx23°$. Now, applying the angle condition, Eq. 4.2: $\theta+90+124+23=540$, $\therefore\theta=303°$.

The locus thus leaves $(-3+j3)$ in the direction 303°. It can then be expected to cross the imaginary axis at a point which can be determined by the Routh–Hurwitz method, and to tend toward the asymptote at 45°. The root locus plot takes the form shown in Fig. 4.5. For low values of K the real root near $s=-1$ and a complex conjugate pair all have significant influence on the response. As K increases the real root has less and less influence, and the system is close to second order. For $K>101.4$ the system is unstable.

Fig. 4.5 Root locus plot for $10K/(s+1)$ $(s+10)(s^2+6s+18)$

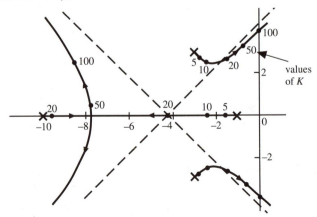

In these examples for low values of K the response is likely to be too sluggish, for high values too oscillatory, but there should be a small range of values of K for which the response is as fast as possible without excessive overshoot. This might be where the dominant roots are at an angle $\alpha = 45°$, corresponding to $\zeta = 0.7$ as in Example 4.1. If the rise time is adequate, and the steady state accuracy for this value of K is within the required specification, then determination of this value of K is all that is required for design. If not, then the loop must be modified in some way such as will be described in the next two sections. It should be noted that the root locus plot identifies only the positions of the roots of the characteristic equation, the values of the exponents p_i in the transient response terms $\sum A_i e^{p_i t}$. The coefficients A_i which determine the relative weighting of the exponential terms are determined by the numerator of the closed loop transfer function $G(s)/(1 + G(s)H(s))$. It is thus necessary to obtain response curves by simulation to confirm the general suitability of the value of K suggested by the root locus plot, and to make adjustment to optimise the response and allow for the effect of this weighting and of the secondary roots, as in Example 4.2.

Comparison of the root locus plots for Examples 4.1 and 4.2 shows that the introduction of a zero into a system has the effect of narrowing the plot, and in particular of pulling the dominant loci towards the left and hence improving stability. Introduction of an additional pole would have the opposite effect of pushing the dominant loci more rapidly into the right half plane and reducing stability.

Frequently it is of interest to know how the positions of the roots of the characteristic equation alter with variation of a parameter other than the gain K. To permit the drawing of plots by the above process, the overall transfer function must first be manipulated to yield an equivalent loop with open loop transfer function $G'(s)H'(s)$ in which the variable parameter appears as a simple multiplying factor.

$$\frac{C(s)}{R(s)} = \frac{G(s)}{1 + G(s)H(s)} = \frac{G'(s)}{1 + G'(s)H'(s)} \qquad [4.6]$$

The characteristic equations $1 + G(s)H(s) = 0$ and $1 + G'(s)H'(s) = 0$ are identical. To determine $G'(s)H'(s)$ the overall transfer function $G(s)/(1 + G(s)H(s))$ is first expanded and written as a ratio of polynomials in s, the denominator is separated into those terms which do not and those which do contain the variable parameter in their coefficients, and both numerator and denominator then divided by the former. With a numerical value assigned to the gain K the plot can then be drawn.

Example 4.4

A unity feedback system has forward loop transfer function of the form $G(s) = K/s(s + 1)(s + \beta)$. Draw a root locus plot to show the variation of the roots of the characteristic equation with β when K has the value 10.

Solution

$$\frac{C(s)}{R(s)} = \frac{K/s(s + 1)(s + \beta)}{1 + (K/s(s + 1)(s + \beta))}$$

$$= \frac{K}{s^3 + (1 + \beta)s^2 + \beta s + K}$$

$$= \frac{K}{(s^3 + s^2 + K) + \beta(s^2 + s)}$$

$$= \frac{K/(s^3 + s^2 + K)}{1 + (\beta s(s + 1)/(s^3 + s^2 + K))} = \frac{G'(s)}{1 + G'(s)H'(s)}$$

$G'(s)H'(s)$ has β as a simple multiplying factor, and a root locus plot can be drawn, provided K is assigned a numerical value.

For $K=10$ there are two zeros at $s=0$ and $s=-1$, and three poles given by the roots of $s^3+s^2+10=0$, the characteristic equation when $\beta=0$. One pole (at least) must be real and can be found by trial and error to be $s=-2.55$. After factorising, $(s+2.55)(s^2-1.55s+3.95)=0$, the other two poles can be calculated to be $0.77\pm j1.83$. The two zeros and the poles are marked on a plot, Fig. 4.6.

Fig. 4.6 Root locus plot for $G(s)H(s)=10/s(s+1)(s+\beta)$

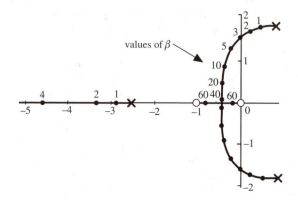

There are three loci, two finishing at the zeros and the third moving to infinity asymptotic to the negative real axis.

There are portions of locus on the real axis from 0 to -1 and from -2.55 to $-\infty$. The crossing of the imaginary axis, from Routh–Hurwitz, is at $s=\pm j1.64$ for a value $\beta=2.7$.

The angle of departure from the complex poles is given by $(68°+46°)-(29°+90°+\theta)=-180°$, i.e. $\theta=175°$. The two dominant loci leave the complex roots in this direction θ, cross the imaginary axis at the points evaluated, break in to the real axis in a normal direction, then separate and move to the two zeros.

For low β the system is unstable, and with increasing β the response will become more damped, and finally overdamped. It must be appreciated that a low value of β implies a large simple lag, since

$$\frac{1}{s+\beta}=\frac{1/\beta}{1+(1/\beta)s}$$

For $\beta=3$ the root positions can be seen to be the same as in Fig. 4.1 for $K=10$, since they represent the same system.

If root locus plots for variation of some parameter β are drawn for a series of values of gain K then a family of curves referred to as a **root contour plot** results which gives information on root positions with variation of two parameters. The starting points for the root contours are points on the root locus for $\beta=0$ with K varying. Figure 4.7 shows a set of contours for the system of Example 4.4. When $\beta=0$, then $G(s)H(s)=K/s^2(s+1)$; there are three poles at $s=0$, $s=0$ and $s=-1$ and no zeros; the three loci are shown as dashed lines.

Following manipulation to enable root contours to be sketched it may be that the order m of the numerator of $G'(s)H'(s)$ exceeds that of the order n of the denominator. If $m>n$ then there will be m loci of which n will start at the poles and the remaining $m-n$ will start at infinity, and all will end at zeros.

Fig. 4.7 Root contour plot for $G(s)H(s) = K/s(s+1)(s+\beta)$

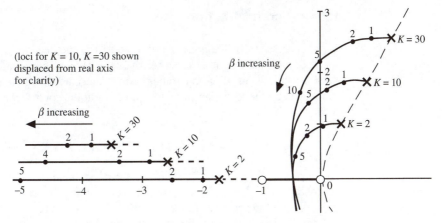

(loci for $K = 10$, $K = 30$ shown displaced from real axis for clarity)

Note that it is not the overall form of a root locus plot which is important, but rather the shape of the dominant loci. Loci well to the left have negligible influence on system behaviour.

4.2 Proportional, integral, derivative and rate feedback control

A desired form of response is generally attained by incorporating some form of controller in a loop, normally in the position shown in Fig. 4.8, namely acting on the error

Fig. 4.8 Feedback system with controller in forward path of loop

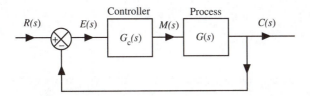

signal to generate an appropriate input signal for the process. The simplest form of controller is a **proportional controller** for which the transfer function $G_c(s)$ is a simple gain term K, which together with the gains of the process and any other elements in the loop makes up the loop gain. A first attempt at design should be to adjust this variable gain K to try to achieve satisfactory closed loop behaviour. It has been seen in Chapter 3 how increase in loop gain reduces the steady state error, but how it also makes the loop more oscillatory, with the open loop harmonic locus moving closer to the critical $(-1, j0)$ point, resulting in reduced gain and phase margins. On the root locus plot the dominant roots move towards the right half plane with increase in gain. Provided that a value of K can be found which positions the dominant roots suitably in the s plane, that simulation confirms the transient behaviour to be suitable (perhaps after fine adjustment), and that with this value of gain the steady state error is acceptable, then the design task is achieved. It is quite likely, however, that for a gain which gives acceptable transient behaviour the steady state error is too large, or the system is too oscillatory whatever the value of gain. The form of the loop transfer function must be modified in some way.

Integral control action

It has been seen that the presence of a factor s in the denominator of $G(s)H(s)$ means that the steady state error is zero when the input signal is constant, which is commonly a system requirement. If an integration term is not inherent in the system then it can be introduced by the inclusion of an integral term in the controller such that $G_c(s)$ has the form

$$\left(k_1 + \frac{k_2}{s}\right), \text{ or } K\left(1 + \frac{1}{T_i s}\right), \text{ or } K\left(1 + \frac{k_i}{s}\right) \dots P + I \text{ action}$$

As long as the input to the controller, the error signal, has any non-zero value then the controller will integrate this to give a rising or falling output which will cause the process output to change. The closed loop will not come to rest until the error has fallen to zero. In the above three forms of transfer function for a proportional plus integral controller the numerical values of the coefficients determine the weighting of the in-tegral action relative to the proportional action, and whether the proportional term can be altered independently or not.

Example 4.5   Investigate the benefits to be gained from the use of a P+I controller for a system with forward loop transfer function $G_p(s) = 1/(1+s)(1+5s)$ and with unity feedback, $H(s) = 1$.

Solution   (i) With proportional control action only, $G_c(s) = K$:

$$\frac{C(s)}{R(s)} = \frac{K/(1+s)(1+5s)}{1 + K/(1+s)(1+5s)} = \frac{K}{5s^2 + 6s + (K+1)}$$

$$= \frac{0.2K}{s^2 + 1.2s + 0.2(K+1)}$$

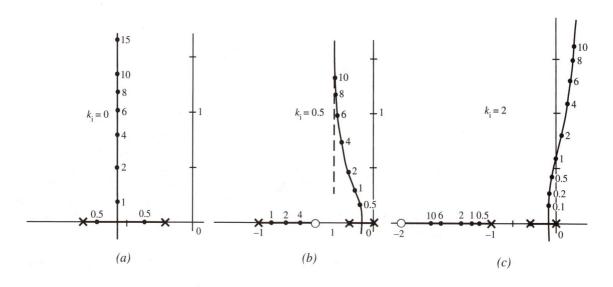

Fig. 4.9   Root locus plots, P+I control, Example 4.5

This is second order, with root locus diagram as in Fig. 4.9(a).

$$\omega_n = \sqrt{\{0.2(K+1)\}}, \quad \text{and} \quad \zeta = 0.6/\sqrt{\{0.2(K+1)\}}$$

For $\zeta = 0.7$, $K = 2.65$.

The steady state error is

$$\lim_{t \to \infty} e(t) = \lim_{s \to 0} sE(s) = \lim_{s \to 0} \frac{sR(s)}{1 + G(s)}$$

which, for a unit step input, $R(s) = 1/s$, is then

$$\frac{1}{1 + \lim_{s \to 0} G(s)} = \frac{1}{1 + K} = 0.274$$

and for a ramp input is $1/\lim_{s \to 0} sG(s) = \infty$.

Fig. 4.10 Unit step responses, corresponding to $\zeta = 0.7$ for dominant roots from Fig. 4.9

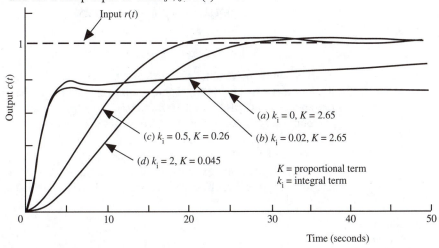

The unit step response is shown as curve (a) in Fig. 4.10. The magnitude of the steady state error could be reduced by increasing K, but this would result in decreased ζ.

(ii) With proportional + integral control, say $G_c(s) = K(1 + k_i/s)$: For a very small value k_i the step response will be almost as above, but then the 'steady state error' will be slowly integrated causing a gentle rise in the output, as shown in curve (b) of Fig. 4.10. This continues until the output reaches unity, at which point the error is zero and the system comes to rest. The integral term also causes a slightly faster rise and increased overshoot. Consider now the root locus plot.

$$G(s)H(s) = \frac{K(1 + k_i/s)}{(1 + s)(1 + 5s)} = \frac{0.2K(s + k_i)}{s(s + 1)(s + 0.2)}$$

There are three poles and one zero, and for $k_i = 0.5$ the root locus diagram is as in Fig. 4.9(b). Comparing this with Fig. 4.9(a) it can be seen that the dominant loci have been pulled towards the right, that for a given value of K the damping factor for the dominant roots is reduced, but that the loop is still stable for all positive values of K. The steady state error for a unit step input is $1/(1 + \lim_{s \to 0} G(s)) = 1/(1 + \infty) = 0$, and for a unit ramp input is $1/\lim_{s \to 0} sG(s) = 1/Kk_i$. The unit step response for $K = 0.26$, the value of K yielding $\zeta = 0.7$ for the dominant roots, is shown as curve (c) in Fig. 4.10. This is more sluggish than curve (a) consistent with a reduction of ω_n for the dominant roots from 0.84 to 0.16 r/s. For the larger value $k_i = 2$ the loci are pulled still farther to the right and the loop now becomes unstable for K greater than a critical value of 1.5, Fig.

4.9(c). The unit step response for $K = 0.045$ ($\zeta = 0.7$ for the dominant roots), curve (d) in Fig. 4.10, is slower than for $k_i = 0.5$ reflecting a further decrease of ω_n to 0.13.

(iii) It should be clear that inclusion of integral action in the controller eliminates the steady state error for a steady input, that the steady state error for a ramp input varies inversely with K, and that inclusion of the integral action without decrease in K results in a more oscillatory response. The designer must determine a suitable combination of values for the two parameters K and k_i. A root contour plot helps in this, and can be sketched by the method described in Section 4.1.

$$\frac{C(s)}{R(s)} = \frac{0.2K(s + k_i)/s(s + 1)(s + 0.2)}{1 + (0.2K(s + k_i)/s(s + 1)(s + 0.2))}$$

$$= \frac{0.2K(s + k_i)}{s^3 + 1.2s^2 + 0.2(K + 1)s + 0.2Kk_i}$$

$$= \frac{0.2K(s + k_i)/s(s^2 + 1.2s + 0.2(K + 1))}{1 + (0.2Kk_i/s(s^2 + 1.2s + 0.2(K + 1)))}$$

There are three poles; one is at $s = 0$ and the other two which are functions of K lie on the root locus of Fig. 4.9(a). There are no zeros and thus the three loci move towards asymptotes at 60°, 180°, and 300° intersecting at $(0 - 0.6 - 0.6)/3 = -0.4$. The root contour plot is shown in Fig. 4.11. Three of the contours which lie superimposed along the negative real axis have been displaced downwards for clarity.

A value of K in the range 2 to 3 together with a value of k_i less than about 0.3 results in $\zeta \approx 0.7$ for the complex roots and should give a response which is not too oscillatory, although the third root on the negative real axis can be expected to significantly

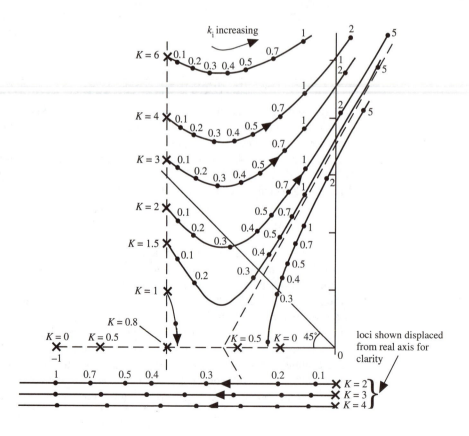

Fig. 4.11 Root contour plot, P + I control, Example 4.5

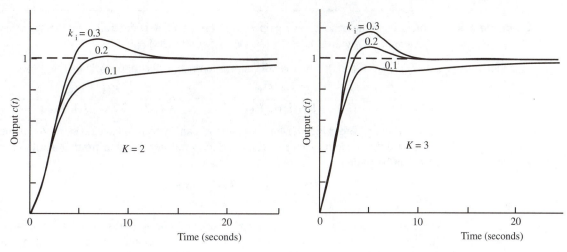

Fig. 4.12   Step response curves for values of K and k_i picked from root contour plot, Fig. 4.11

influence the behaviour. Figure 4.12 shows unit step response curves obtained by simulation for this range of parameter values. $K = 3$ and $k_i = 0.2$ gives a response with perhaps the best combination of small rise time, overshoot, and settling time, and also gives a smaller steady state error with a ramp input than for $K = 2$. Comparison of the six curves suggests that $K = 4$ and $k_i \approx 0.1$ would be worth looking at. In fact with $K = 4$ the best value of k_i is around 0.15 giving a smaller rise time, longer settling time, similar overshoot, and small undershoot as compared with the best of the six curves in Fig. 4.12. The nature of the application would dictate the relative importance of the different aspects of the behaviour and hence which combination of parameters would be deemed 'optimum'. It should be noted that the settling time attained for the loop is about half of that for the process with its dominant time constant of 5 second.

Derivative control action

The controller may alternatively incorporate derivative action, so that to the signal proportional to error is added a signal proportional to the derivative of the error. The transfer function then has the form

$$(k_1 + k_3 s) \text{ or } K(1 + k_d s) \dots \text{P} + \text{D action}$$

If the rate of change of error is large then a large overshoot or undershoot can be expected, and derivative action anticipates this. It reduces the oscillation by *decreasing* the manipulated variable $m(t)$ when $c(t)$ is moving *towards* the desired value (since $e(t)$ and $\dot{e}(t)$ then have opposite signs) and *increasing* $m(t)$ when $c(t)$ is moving *away from* the desired value (since $e(t)$ and $\dot{e}(t)$ have the same sign). Derivative action has no effect on the steady state error with a constant input, since $\dot{e}(t)$ is then zero. A disadvantage with real systems is that differentiation will amplify any high frequency noise which may be present in the signal, which is potentially damaging to the input elements of the process being controlled (even though the noise will be attenuated in its passage round the loop). The above transfer function, with numerator polynomial of higher order than denominator, is not physically realisable, so in practice there will also be a small associated lag.

Example 4.6   Investigate the benefit to be gained from use of a P + D controller for the system of Example 4.5, with $G_p(s) = 1/(1 + s)(1 + 5s)$ and $H(s) = 1$.

Solution   Let $G_c(s) = K(1 + k_d s)$. The overall (closed loop) transfer function is then

$$\frac{C(s)}{R(s)} = \frac{K(1 + k_d s)}{(1 + s)(1 + 5s) + K(1 + k_d s)} = \frac{0.2K(1 + k_d s)}{s^2 + (1.2 + 0.2Kk_d)s + 0.2(K + 1)}$$

This is a second order transfer function, and k_d can be seen to have no effect on ω_n but to increase the damping factor ζ. The open loop transfer function from which the root locus can be plotted is

$$G(s)H(s) = \frac{K(1 + k_d s)}{(1 + s)(1 + 5s)} = \frac{0.2K/k_d(s + 1/k_d)}{(s + 1)(s + 0.2)}$$

There are two poles at $s = -1$ and $s = -0.2$ and one zero at $s = -1/k_d$, and hence there are two loci, one ending at the zero and the other moving to $-\infty$ along the negative real axis. Figure 4.13 shows the root loci for three values of k_d. (Note that the loci shown between $-\infty$ and -1.11 are for $k_d = 0.9$, those for $k_d = 0.5$ and 0.25 having been omitted in this region for clarity.) Comparison with the locus for $k_d = 0$ shows that the derivative term pulls the loci to the left, and that for a given value of gain K the damping factor ζ of the roots is increased. P + D control permits the use of higher gains with a resultant increase in speed of response and decrease in steady state error, whilst avoiding a more oscillatory response. The root locus plot, Fig. 4.13, indicates

Fig. 4.13   Root locus plots, P + D control, Example 4.6

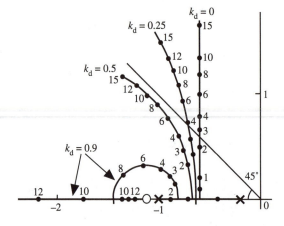

that a value of $k_d \approx 0.4$ with a gain K of 10 or greater might give the 'best' response. Figure 4.14 shows the unit step response for a few of the root positions marked on the root locus plot.

Three term controller

The advantages offered by integral and derivative action are complementary, hence it can be useful to incorporate both in a **three term controller** or **PID controller** with transfer function $G_c(s)$ of the form

$$(k_1 + k_2/s + k_3 s) \text{ or } K(1 + 1/T_i s + T_d s)$$

Fig. 4.14 Step responses, P + D control, Example 4.6

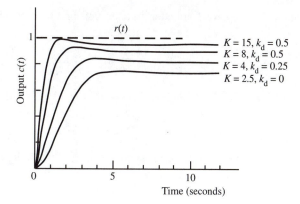

$r(t)$

$K = 15, k_d = 0.5$
$K = 8, k_d = 0.5$
$K = 4, k_d = 0.25$
$K = 2.5, k_d = 0$

Output $c(t)$

Time (seconds)

The approach to selection of suitable values of the three parameters can be with the aid of root contour plots and simulation, as above. Industrially, where mathematical models can only be approximations to the plant behaviour, optimum parameter settings are determined by empirical methods for tuning the controller, the best known being based on recommendations by Ziegler and Nichols.

Rate feedback, or negative velocity feedback

The anticipatory action of derivative control can be achieved by differentiating the output signal and feeding it back by a minor loop as in Fig. 4.15. Except when $r(t)$ is

Fig. 4.15 Block diagram for rate feedback

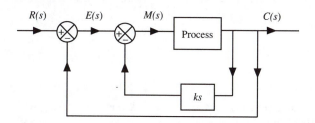

$R(s)$ $E(s)$ $M(s)$ Process $C(s)$

ks

changing, as at $t = 0$ when a step input function is applied or when $r(t)$ is a ramp function, then differentiation of $c(t)$ prior to subtraction from $e(t)$ is equivalent to differentiation of $e(t)$. If the system is a positional servomechanism where $c(t)$ is a shaft angular position (or proportional to it) then a voltage proportional to $\dot{c}(t)$ can readily be obtained from a tachogenerator mounted on the motor shaft. This avoids the problems of amplification of any high frequency noise component of a signal inherent in the process of differentiation, whether effected by analogue or digital means. As will be illustrated by the next example, the characteristic equation is the same as for P + D control, but the numerator of the overall transfer function is different, resulting in a dynamic response which is similar but not identical.

Example 4.7   Investigate the effect of negative velocity feedback on a servomechanism in which the 'process' of Fig. 4.15 is an amplifier and motor with transfer function $K/s(1 + s)$. How does this compare with the use of derivative action in the forward loop?

Solution   The overall transfer function will first be derived. The inner loop can be replaced by a block with transfer function

$$\frac{K/s(s+1)}{1+Kk/(s+1)} = \frac{K}{s^2 + s + Kks}$$

and then the main loop reduced to a single block with transfer function

$$\frac{C(s)}{R(s)} = \frac{K/(s^2 + s + Kks)}{1 + K/(s^2 + s + Kks)} = \frac{K}{s^2 + (1 + Kk)s + K} \qquad [1]$$

Note that Fig. 4.15 is equivalent to a single loop with transfer function $(1 + ks)$ in the feedback path, which also reduces to Eq. [1].

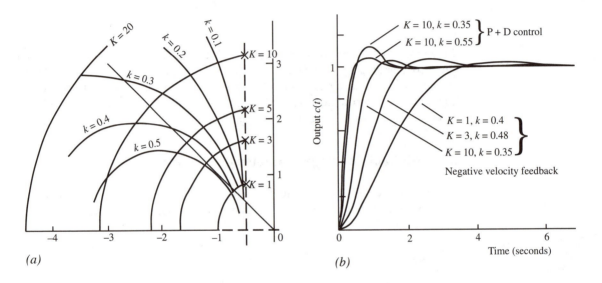

(a)   *(b)*

Fig. 4.16   Root contour plot and step response curves, Example 4.7

It is a second order system with $\omega_n = \sqrt{K}$ and ζ increasing with increase in K. The root contours, Fig. 4.16(a), follow circular arcs since ω_n is independent of k; it is convenient to draw also loci of constant k. This plot indicates combinations of K and k which would yield any desired value of ζ for the dominant roots. Figure 4.16(b) shows the unit step response for $K = 1$, 3 and 10 with $k = 0.4$, 0.48 and 0.35 respectively chosen so that $\zeta \approx 0.7$. It can be seen that as the gain K is increased a faster response without increased overshoot can be achieved. There will be a limit to how far K can be increased since the amplifier may reach saturation or the voltage may become too great for the motor; also the assumption of linearity may not remain valid.

The alternative system with a P + D term $(1 + ks)$ in the forward loop has overall transfer function

$$\frac{C(s)}{R(s)} = \frac{K(1 + ks)/s(s+1)}{1 + K(1 + ks)/s(s+1)} = \frac{K(1 + ks)}{s^2 + (1 + Kk)s + K} \qquad [2]$$

It can be seen that the characteristic equation is identical to that for Eq. [1], and hence the root contour plot Fig. 4.16(a) still applies. The step response, as typified by the curve for $K = 10$ and $k = 0.35$, is however markedly different, with a greater slope at

$t = 0$ and a greater overshoot, a consequence of the different numerators in Eqs [1] and [2]. This can be understood qualitatively by appreciating that at $t = 0$ the error changes instantaneously from zero to unity (prior to the subsequent finite rate decrease), which contributes to the output of a P + D controller but not a rate feedback loop. Analytically, with rate feedback and a unit step input:

$$C(s) = \frac{10}{s(s^2 + 4.5s + 10)}$$

$$= \frac{1}{s} - \frac{s + 4.5}{s^2 + 4.5s + 10}$$

$$= \frac{1}{s} - \frac{s + 2.25}{(s + 2.25)^2 + (2.222)^2} - \frac{1.012(2.222)}{(s + 2.25)^2 + (2.222)^2}$$

$$\therefore \ c(t) = 1 - e^{-2.25t}(\cos 2.222t + 1.012 \sin 2.222t) \tag{3}$$

With a P + D controller

$$C(s) = \frac{10(1 + 0.35s)}{s(s^2 + 4.5s + 10)}$$

$$= \frac{1}{s} - \frac{s + 1}{s^2 + 4.5s + 10}$$

$$= \frac{1}{s} - \frac{s + 2.25}{(s + 2.25)^2 + (2.222)^2} + \frac{0.563(2.222)}{(s + 2.25)^2 + (2.222)^2}$$

$$\therefore \ c(t) = 1 - e^{-2.25t}(\cos 2.222t - 0.563 \sin 2.222t) \tag{4}$$

It is the different weightings of the cosine and sine terms in the two Eqs [3] and [4] which causes the responses to differ. This example illustrates the need to exercise caution in the use of root locus plots, and the need to supplement them by simulation. Increase of k to about 0.55 is required to reduce the overshoot to that with rate feedback, and this represents a large change in root position.

4.3 Phase compensation

An alternative approach to system improvement is to insert in the forward loop, in series with the process, a passive element with transfer function of the form $(1 + \tau_1 s)/(1 + \tau_2 s)$, i.e. an element comprising both a simple lead term and a simple lag term. If $\tau_1 > \tau_2$ then the lead will dominate and it is referred to as a **phase lead element**, and if $\tau_1 < \tau_2$ then the lag will dominate, and it is a **phase lag element**. Figure 4.17 shows how this transfer function can be achieved by a d.c. electrical network; there are mechanical equivalents utilising springs and dampers. Design is conventionally carried out in the frequency domain using the Bode plot, and this method will be described and illustrated by examples. It should by now be clearly understood that for any given closed loop system increase in loop gain will improve steady state accuracy and reduce stability, and vice versa. Unless the steady state error and the stability requirements can be simultaneously satisfied by some value of gain K, then if K is set to give the required accuracy, $G(j\omega)$ will be too close to the critical $(-1, j0)$ point on the Nyquist plot, or if K is set to give the required phase margin, then the steady state error will be too large.

Fig. 4.17 Phase lead and phase lag networks

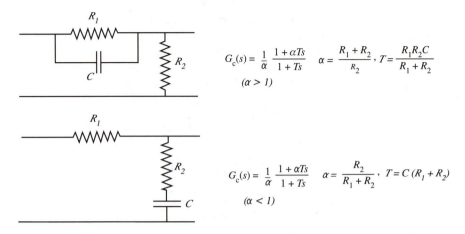

$$G_c(s) = \frac{1}{\alpha} \frac{1 + \alpha Ts}{1 + Ts} \quad \alpha = \frac{R_1 + R_2}{R_2}, \; T = \frac{R_1 R_2 C}{R_1 + R_2}$$

$$(\alpha > 1)$$

$$G_c(s) = \frac{1}{\alpha} \frac{1 + \alpha Ts}{1 + Ts} \quad \alpha = \frac{R_2}{R_1 + R_2}, \; T = C(R_1 + R_2)$$

$$(\alpha < 1)$$

The principle of phase compensation is one of modifying the harmonic locus locally so that both requirements are simultaneously achieved. Three methods are available: phase lead compensation, phase lag compensation, and a combination of the two.

Figures 4.18(a) and (b) show the nature of the harmonic response of a phase lead and a phase lag element written in the form $(1 + \alpha Ts)/(1 + Ts)$, with $\alpha > 1$ and $\alpha < 1$

Fig. 4.18 Bode plots for: (*a*) phase lead element; (*b*) phase lag element

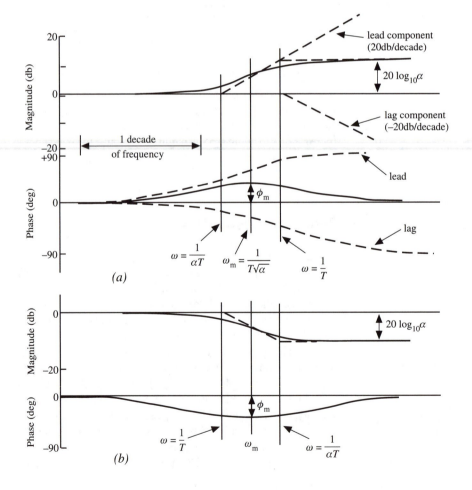

respectively. For a phase lead element the magnitude is 0 db for very low frequencies and 20 $\log_{10}\alpha$ db at very high frequencies, and the straight line approximation has break points at $\omega = 1/\alpha T$ and $\omega = 1/T$. The phase is 0° at very low frequencies, and as frequency progressively increases the phase starts to increase due to the effect of the lead term, then reaches a maximum and falls again as the lag term exerts its influence. Finally the phase falls back to zero for very high frequencies. Owing to symmetry the maximum phase lead ϕ_m is at a frequency ω_m midway between the two break points on the logarithmic scale.

$$\log \omega_m = 1/2[\log_{10}(1/\alpha T) + \log_{10}(1/T)] = 1/2 \log_{10}(1/\alpha T^2)$$

$$\therefore \ \omega_m = 1/T\sqrt{\alpha} \tag{4.7}$$

Also, $\phi_m = \tan^{-1}(\omega_m \alpha T) - \tan^{-1}(\omega_m T)$

$$\therefore \ \tan \phi_m = \frac{\omega_m \alpha T - \omega_m T}{1 + (\omega_m \alpha T)(\omega_m T)} = \frac{\sqrt{\alpha} - 1/\sqrt{\alpha}}{2} = \frac{\alpha - 1}{2\sqrt{\alpha}}$$

or

$$\sin \phi_m = \frac{\alpha - 1}{\alpha + 1} \tag{4.8}$$

The maximum possible value of ϕ_m is 90° to which it would tend as $\alpha \to \infty$. For a phase lag element the magnitude and phase plots are the inverse of those for a phase lead element, mirror images about the 0 db and 0 degree lines respectively, Fig. 4.18(b). As frequency increases the influence of the lag term is felt before that of the lead.

Phase lead compensation

Figure 4.19 illustrates the manner in which the phase margin of a closed loop system is improved by the introduction of a phase lead element. It shows the magnitude and phase curves for an uncompensated system $G(s)$ with the gain assumed to be set to a value at which the steady state error does not exceed the maximum permissible. This results in a

Fig. 4.19 Illustration of phase lead compensation

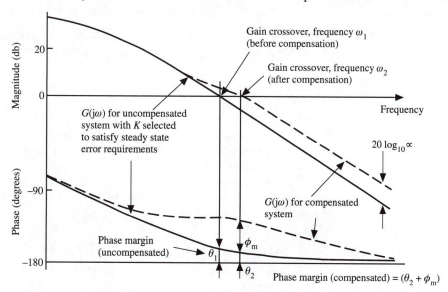

phase margin θ_1, assumed to be too small, at the gain crossover frequency ω_1. The principle of phase lead compensation is to select a value of α which results in a value ϕ_m sufficient to increase the phase margin to at least the minimum specified value, and to select T to place ϕ_m at the best position along the frequency axis. The phase lead element also modifies the magnitude, and hence increases the gain crossover frequency. The design aim is to place ϕ_m at the new gain crossover frequency ω_2, which is achieved by selecting ω_m to be the frequency where the uncompensated gain has the value $-1/2$ (20 $\log_{10}\alpha$). In choosing α allowance must be made for the fact that the phase of the uncompensated system at ω_2 has changed from that at ω_1. Finally, the gain K must be increased by a factor α to compensate for the attenuation $1/\alpha$ inherent with the electrical circuit of Fig. 4.17(a) (where the gain is unity or zero db at high frequencies, and $1/\alpha$ or $-20 \log_{10} \alpha$ db at low frequencies). The steps of the design process, to be followed in the next example, are thus:

1. Calculate the value of gain K which will satisfy the steady state error requirements, and draw on a Bode plot the magnitude and phase curves of the system with this gain.
2. Measure the phase margin, subtract from the required minimum value, and add 4 or 5 degrees (an estimate of $\theta_1 - \theta_2$) to determine the required value of ϕ_m.
3. Calculate α using Eq. 4.8.
4. Calculate 20 $\log_{10} \alpha$, the phase lead element high frequency magnitude, and estimate the frequency at which the uncompensated system has magnitude $-1/2(20 \log_{10} \alpha)$. Call this ω_m.
5. Calculate the phase lead element break points $\omega_m/\sqrt{\alpha}$ and $\omega_m\sqrt{\alpha}$, and draw the magnitude and phase curves for the phase lead element.
6. Add these to the curves for the system to yield the frequency response of the compensated system, and check that the phase margin has at least the required value. If not, return to step 2 and repeat.

Example 4.8   Design a phase lead series compensation network for a unity feedback system with forward loop transfer function $K/s(1+s)$ to meet performance requirements of a phase margin of 45°minimum and a velocity error constant of 25 second^{-1}.

Solution   With a step input the steady state error will be zero as a consequence of the factor s on the denominator of $G(s)$. With a unit ramp input the steady state error will be

$$1/\lim_{s\to 0} sG(s) = 1/K_v = 1/25 = 0.04$$

Now $K_v = \lim_{s\to 0} s[K/s(1+s)] = K$, hence to meet the steady state error requirements K must have the value 25.

Draw the magnitude and phase curves for $G(s) = 25/s(1+s)$ on a Bode plot using straight line approximations, Fig. 4.20. In the region of the gain crossover frequency, around 5 r/s, draw the curves more accurately (either by including the corrections described in Section 2.4, or by calculation for selected values of frequency, or by obtaining the information from control analysis software). The phase margin can now be measured (or calculated) to be about 11°, at a gain crossover frequency of close to 5 r/s.

$$\text{Magnitude} = 20 \log_{10} [25/5\sqrt{(1 + 5^2)}] = -0.17$$
$$\text{Phase} = -90° - \tan^{-1} 5 = -168.7° \qquad \therefore \ \text{Phase margin} \approx 11.3°$$

Fig. 4.20 Phase lead compensation, Example 15.8

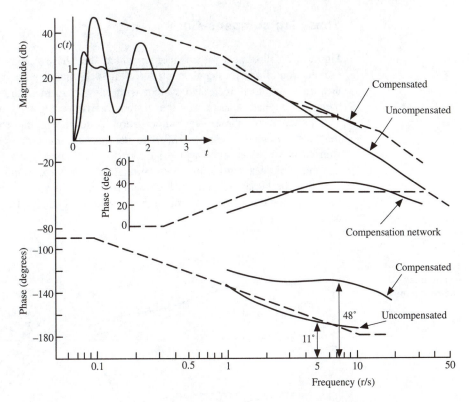

The minimum phase lead required from the compensation is thus 34°. Design for $\phi_m = 40°$ to make allowance for the decrease in phase which will accompany the increase in gain crossover frequency:

$$\therefore \ \frac{\alpha - 1}{\alpha + 1} = \sin 40° = 0.643 \qquad \therefore \ \alpha = \frac{1 + 0.643}{1 - 0.643} = 4.6$$

The phase lead element will thus have transfer function $(1 + 4.6Ts)/(1 + Ts)$ where T must now be chosen to place ϕ_m at a suitable frequency. The high frequency gain is $20 \log_{10} 4.6 = 13.26$ db. Place ϕ_m at the frequency where $|G(j\omega)| = -0.5(13.26) = -6.63$ db, i.e. at $\omega = 7.3$ r/s. The corner frequencies of the phase lead element are thus $7.3/\sqrt{4.6} = 3.4$ r/s and $7.3\sqrt{4.6} = 15.6$ r/s, corresponding to time constants of $1/3.4 = 0.30$ second and $1/15.6 = 0.064$ second. The transfer functions of the compensation network and the process (whose gain must be increased by the factor 4.6 to compensate for the attenuation of the compensation network) are

$$G_c(s) = \frac{1}{4.6} \frac{1 + 0.30s}{1 + 0.064s} \quad \text{and} \quad G(s) = \frac{115}{s(1 + s)}$$

The phase margin is now 48°, at the increased gain crossover frequency of 7.3 r/s. This comfortably exceeds the 45° because of the small slope on the uncompensated phase curve and the generous allowance of 6°. The inset in Fig. 4.20 shows the step responses before and after compensation. Before compensation the response is very oscillatory; after compensation the peak overshoot is reduced to 25%, the rise time is reduced, and the settling time reduced almost to one second.

Phase lag compensation

Figure 4.21 illustrates the manner in which the phase margin is improved by the introduction of a phase lag element. As in Fig. 4.19 the magnitude and phase curves are drawn for the uncompensated system with gain set to meet the steady state requirements. The phase margin for the system drawn is negative, indicating instability. The principle of phase lag compensation is to utilise the characteristic of high frequency attenuation (Fig. 4.18(b)) to decrease the magnitude in the region of the gain crossover, thus reducing the gain crossover frequency and increasing the phase margin, whilst leaving the phase curve relatively unaltered at this frequency. It can be seen that the lag associated with the compensation element is arranged to occur in a frequency region well below the gain crossover frequency. The steps of the design process are thus:

Fig. 4.21   Illustration of phase lag compensation

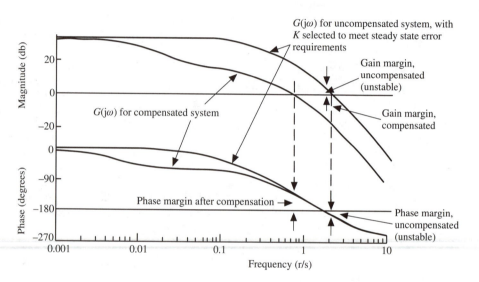

1. Calculate the value of gain K which will satisfy the steady state error requirements, and draw on a Bode plot the magnitude and phase curves of the system with this gain.
2. Measure the frequency corresponding to the required phase margin plus about 5°; call this ω_c. Measure the magnitude of $G(j\omega)$ at this frequency; call this G_c. The phase lag element must introduce an attenuation of G_c so that the magnitude curve of the compensated system passes through 0 db near the frequency ω_c.
3. Calculate α, from the equation $20 \log_{10} \alpha = G_c$.
4. Select T so that the higher corner frequency is one decade below ω_c, i.e. $1/\alpha T = 0.1\omega_c$, in which case the lag introduced at ω_c is around 5°, which has been allowed for.
5. Draw the magnitude and phase curves for the phase lag element with corner frequencies $1/T$ and $1/\alpha T$.
6. Add these to the curves for the system to yield the frequency response of the compensated system, and check that the phase margin is as required.

Example 4.9   Design phase lag series compensation for a unity feedback system with forward loop transfer function $K/(1 + 5s)(1 + s)(1 + 0.5s)$ in order to achieve a phase margin of at least $45°$ and a positional error constant of 40.

Solution $K_p = \lim_{s \to 0} [K/(1 + 5s)(1 + s)(1 + 0.5s)] = K$. Hence the gain K must have the value 40, and the steady state error for a unit step input will be $1/(1 + K_p) = 1/41$, which is about 2.5%. The magnitude and phase curves are those shown in Fig. 4.21. It would suffice to draw straight line approximations and calculate accurately only values in the frequency range 0.5 to 3 r/s. The phase margin is $-18°$, i.e. the system is unstable with $K = 40$. The phase margin would be $45 + 5 = 50°$ if the gain crossover were at a frequency ω_c of about 0.7 r/s. The magnitude at this frequency is 18.6 db, and this therefore is the high frequency attenuation required of the phase lag compensation element to ensure that gain crossover is at ω_c:

$$20 \log_{10} \alpha = -18.6 \qquad \therefore \ \alpha = 0.12$$

The corner frequencies of the compensation element (Fig. 4.18(b)) are $1/T$ and $1/\alpha T$. Choose T so that the higher of these is at one tenth of ω_c:

$$1/\alpha T = 0.1(0.7) \qquad \therefore \ T = 119, \ \text{say } 120, \ \text{and } \alpha T = 14, \ \text{say.}$$

The transfer functions of the compensation network and the process are therefore:

$$G_c(s) = \frac{1 + 14s}{1 + 120s} \ \text{and} \ G(s) = \frac{40}{(1 + 5s)(1 + s)(1 + 0.5s)}$$

The magnitude and phase curves for $G_c(s)G(s)$, the compensated open loop response, are shown in Fig. 4.21. The $5°$ allowance was appropriate, and the phase margin after compensation is $45°$.

Whether one of these methods of system compensation is likely to be successful depends to a large extent on the shape of the phase plot for the uncompensated system. If the slope of the phase curve is large at the gain crossover frequency, then attempts at phase lead compensation are likely to fail since the amount of phase lost due to the unavoidable increase in gain crossover frequency can largely negate the phase lead gained from the compensation network. If the process contains a double integration, a factor s^2 on the denominator, then the phase curve will lie largely or completely below the $-180°$ level, and phase lag compensation cannot be effected. The illustrations above show that the two forms of compensation affect the dynamic behaviour in different ways. Phase lead compensation leads to an increase in gain crossover frequency, which is accompanied by increases in ω_p and the bandwidth (which can be understood by visualising movement of the harmonic locus on the Nichols chart). As a consequence, in addition to making the response less oscillatory the rise time is reduced (faster response) but the penalty of increased bandwidth is a greater susceptibility to noise. The phase lead element acts as a high pass filter. Attempts to introduce a phase lead of greater than $55°$ or so may also well lead to a control signal $m(t)$ which is too large for the process. In contrast phase lag compensation reduces the gain crossover frequency, and with it ω_p and bandwidth, and hence in addition to making the response less oscillatory it increases the rise time (slower response with long settling time). A compromise can be achieved, with little change in gain crossover frequency and bandwidth, by means of **lag–lead compensation** using a series compensation element with transfer function

$$G_c(s) = \frac{1 + \alpha T_1 s}{1 + T_1 s} \frac{1 + \beta T_2 s}{1 + T_2 s} \text{ with } \alpha < 1, \ \beta > 1, \ \text{and} \ T_1 > T_2$$

Alternatives to the use of the Bode plot for design are the use of the Nichols chart, and root locus diagrams. In all cases simulation would be used to ensure that the transient behaviour was acceptable, and perhaps to fine tune the design to achieve a better compromise between different response characteristics. It may prove more suitable overall, say, to arrange ϕ_m with phase lead compensation to be somewhat offset from the gain crossover frequency.

4.4   Introduction to more advanced concepts and techniques

These four chapters have treated the most basic concepts of feedback systems, and classical linear control theory which permits fundamental analysis. Real systems are often more complex, may exhibit significant non-linearity, may have signals which are not deterministic or not continuous, may require some different form of control, or in some other way not be suitable for analysis entirely by the methods which have been described. Brief outlines are now given of such areas of control engineering to give the reader an awareness of them. Application of any techniques in these areas will require study of more comprehensive control engineering text books, and for this purpose and also to amplify the basics covered in this book the Bibliography following this chapter lists books which will be found useful.

State space representation

This is the basis of 'modern control theory' and dates from the 1960s. Instead of a dynamic system or component being described by an nth order differential equation (written as an nth order transfer function) the component is described by n first order differential equations, each of the form $\dot{x}(t) = ax(t) + bu(t)$. The n equations are written in compact form as the matrix state equation

$$\dot{x}(t) = \mathbf{A}x(t) + \mathbf{B}u(t) \tag{4.9}$$

and the output is defined by the matrix output equation

$$y(t) = \mathbf{C}x(t) + \mathbf{D}u(t) \tag{4.10}$$

The bold symbols represent vectors and matrices. The vector $x(t)$ containing the n variables $x_1(t), x_2(t), \ldots, x_n(t)$ describes the state of the system and is referred to as the **state vector**; $\dot{x}(t)$ is also n dimensional. $u(t)$ and $y(t)$ are the input and output vectors and have orders equal to the numbers of inputs and outputs respectively. $\mathbf{A}$, $\mathbf{B}$, $\mathbf{C}$ and $\mathbf{D}$ are matrices where $\mathbf{A}$ has order $(n \times n)$ and $\mathbf{B}$, $\mathbf{C}$ and $\mathbf{D}$ must have the correct orders to satisfy the rules of matrix multiplication. There is no single unique set of variables which form the state vector; some would have more direct physical meaning than others, and some might be easier for analysis. The characteristic equation is given by $|s\mathbf{I} - \mathbf{A}| = 0$, where $\mathbf{I}$ is a unit vector, and a generalised matrix transfer function $\mathbf{G}(s)$

which defines all the output–input relationships of the form $Y_i(s)/U_j(s)$ is given in terms of the four matrices in Eqs 4.9 and 4.10 by

$$G(s) = \mathbf{C}(s\mathbf{I} - \mathbf{A})^{-1}\mathbf{B} + \mathbf{D}$$

The main advantages of a state space description over a transfer function are that it can describe a system with more than one input and output (a multi-input–multi-output system), and that it can be used to handle complex systems without markedly greater difficulty than simple systems. Analysis makes use of standard computer algorithms for solving first order equations and for matrix manipulation. A disadvantage, and the reason why it is not covered in detail in this book, is that it does not impart such a clear physical appreciation of the nature of system behaviour as classical theory does. It also requires a stronger mathematical background. Matrix methods are likewise used extensively in dynamics for complex systems.

Non-linear systems

As explained in Section 2.1, a linear system is one for which the principle of super-position is valid, and this requires the differential equations to have coefficients which are not functions of the variables or their derivatives, and to contain no powers or products of the variables or derivatives. Linear analysis is the most straightforward, hence a sensible first approach with a non-linear system is to try to linearise the equations. If this can be done then the linear response should be close to the actual for small deviations from the datum, and a reasonable guide for larger perturbations. Certain non-linearities such as dead zone or hysteresis have their greatest effect when changes to the system state are small and hence cannot be approximated by linearis-ation. Three approaches to analysis of non-linear systems will be outlined.

Non-linearities of any form can be investigated by **simulation** to study the changes to system response which result from parameter variations or changes of model form. Used by itself this is a somewhat trial and error method which does not clearly show causal effects, and hence it is best employed to confirm conclusions suggested by some form of analysis.

The **phase plane technique** is a graphical approach which can be applied to any second order system, either linear or non-linear, but not to a higher order system except insofar as it may be possible to approximate it by a second order model. It will be illustrated by reference to a feedback system with relay control, also referred to as bang-bang control or on–off control. Such control is simple and cheap and is widely used for heating/cooling systems and simple servomechanisms. Figure 4.22(a) shows the block diagram for a relay controlled servomechanism with an ideal relay controlling a d.c. motor with $G(s) = 1/s(1 + \tau s)$. The phase plane plot has axes of output $c(t)$ and its derivative $\dot{c}(t)$, or more usually $e(t)$ and $\dot{e}(t)$ as in Fig. 4.22(b). On it are drawn trajectories showing how the state of the system changes with time if released from different initial states in the absence of any forcing function. Note that since $e(t) = r(t) - c(t)$, if $r(t) = $ constant then $e(t)$ is the same as $-c(t)$ with a shift of datum, and $\dot{e}(t) = -\dot{c}(t)$. The motor input is either $+K$ when $e(t) > 0$, in which case the motor torque has a certain positive value and one of the trajectories of the form P_1P_2 or Q_1Q_2 is followed (that which passes through the initial state), or the input is $-K$ when $e(t) < 0$, which produces a negative torque and movement along one of the trajectories of the form N_1N_2. Upon reaching $e(t) = 0$ along a trajectory from the right the relay will

Fig. 4.22 Phase plane
plots for relay
controlled
servomechanism: (*a*),
(*b*) ideal relay; (*c*), (*d*)
relay with dead zone
$\pm \delta$

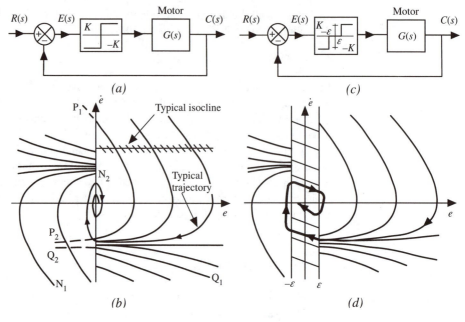

switch, the torque will change direction, and a new trajectory will be followed until
$e(t) = 0$ again and the relay switches back. With an ideal relay switching will occur
more and more rapidly, but the system never comes to rest. In reality a relay has some
dead zone, Fig. 4.22(c), which introduces to the phase plane plot a narrow band where
motor torque is zero and through which the system 'coasts', Fig. 4.22(d). Once the
$\dot{e}(t) = 0$ axis is reached within the dead zone, the system will come to rest with an error
in the range $-\varepsilon \leq e(t) \leq \varepsilon$. The phase portrait can be drawn by evaluating $\dot{e}(t)$ as a
function of $e(t)$, if simple, or by a graphical approach called the method of isoclines.
When the equation of an isocline, a line crossed by all trajectories at a given slope, is
found, then a set of isoclines can be drawn with short lines indicating the slope (for Fig.
4.22(b) they are horizontal lines and a typical one is shown). Starting from any given
initial state, a trajectory can then be drawn to cross all the isoclines at the correct slope.

The **describing function method** is a frequency domain approach applicable to a
system of any order provided there is a single non-linearity, or that all the non-linearities
can be lumped into one. If a sine wave is applied at the input to a non-linearity, then the
output, although periodic, will not be sinusoidal, but made up of a sine wave at the
input frequency (the fundamental component) and a number of harmonics of different
amplitudes at integer multiples of the fundamental frequency. When the non-linearity is
in a closed loop then the higher frequency components of the output signal will be
attenuated more than the fundamental by the lags around the loop so that the signal
returning to the non-linearity is little different to what it would be if the output had
simply been the fundamental component. The non-linearity can therefore be closely
represented by a function N, referred to as a describing function, which defines the
amplitude and phase of the fundamental relative to an input sinusoid, and is thus
equivalent to a harmonic transfer function $G(j\omega)$. N, which is evaluated analytically by
Fourier decomposition of the output signal, is a function of input amplitude A, and will
often simply be a variable gain; if there is also phase shift then N will be complex; for
certain non-linearities N is a function of both A and frequency ω.

A characteristic of non-linearities is that they can cause small amplitude steady state oscillations, referred to as **limit cycles**, which persist even when there is no forcing function. The describing function method enables one to determine whether limit cycles are likely to occur, and if so their amplitude and frequency, and how they might be reduced or eliminated. Consider a unity feedback system with N and $G(s)$ in the forward loop. When $R(s) = 0$ then $C(s) = NG(s)E(s) = -NG(s)C(s)$ $\therefore C(s)(1 + NG(s)) = 0$. Now $C(s)$ can only be non-zero and oscillations sustained if $(1 + NG(s))$ is zero, i.e. if $G(s) = -1/N$. The approach is to plot $G(j\omega)$ and $-1/N$ on a polar diagram (Nyquist plot) and study the intersection, if any. Figure 4.23(a) shows these curves for a third

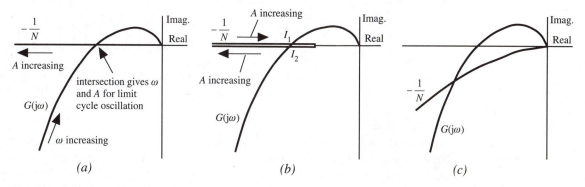

Fig. 4.23 Illustration of use of describing function for relay controlled feedback system: (*a*) ideal relay; (*b*) relay with dead zone; (*c*) relay with hysteresis

order process with an ideal relay, for which the output is a square wave of amplitude K and $N = 4K/\pi A$ which is a gain which varies inversely with A. It can be seen that the curves intersect, which indicates that a limit cycle can be expected of frequency given by the value of ω on the $G(j\omega)$ curve at intersection, and amplitude at input to the relay given by the value of A on the $-1/N$ curve at intersection. The nature of the limit cycle can be explained by reference to the Nyquist criterion which states that the system will be unstable if the plot of $NG(j\omega)$ encloses the critical $(-1, j0)$ point, i.e. if $G(j\omega)$ encloses the $-1/N$ point. If as a consequence of some disturbance there is an oscillation of very small amplitude A, then the corresponding critical point $-1/N$ will lie within the $G(j\omega)$ curve, the system will be unstable and hence the amplitude A will increase and the $-1/N$ point will move out towards the $G(j\omega)$ curve. Conversely, if there is an oscillation of large amplitude A, then the critical point $-1/N$ will lie outside $G(j\omega)$, the system will be stable, and hence the amplitude A will decrease and $-1/N$ will move in towards $G(j\omega)$. The system thus approaches a steady state with limit cycle oscillations when conditions are such that $G(j\omega)$ and $-1/N$ intersect. If the relay has a dead zone d then $N = (4K/\pi A)\sqrt{(1 - (d/A)^2)}$ for $A \geq d$ which reaches a maximum value for $A = (\sqrt{2})d$. The $-1/N$ locus intersects $G(j\omega)$ either twice (Fig. 4.23(b)) or not at all. One of the intersections I_2 represents a stable limit cycle as described above, the other I_1 an unstable limit cycle of the same frequency from which state a slight disturbance would cause the oscillation either to decay to zero or to increase to that of the stable limit cycle. If the gain of $G(j\omega)$ is reduced until the curve no longer intersects with $-1/N$ then the limit cycle would be eliminated. For a relay with hysteresis $N = (4K/\pi A) \angle \sin^{-1}(h/A)$ and $-1/N$ no longer lies along the negative real axis (Fig. 4.23(c)).

Sampled data systems

Digital processors, when used as controllers, operate not on continuous signals, but on discrete representations of them. By means of a sampler, a switch which closes for a very short time every T seconds, a continuous signal $e(t)$ can be transformed into a train of amplitude modulated pulses $e*(t)$, Fig. 4.24. Most of the techniques for continuous

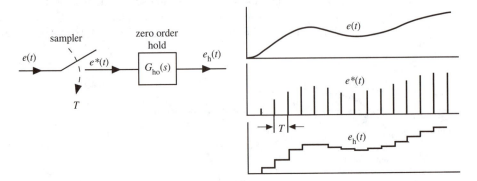

system analysis have been adapted to handle sampled data systems, in particular by introducing a variable $z = e^{sT}$ in place of the Laplace operator s. A continuous function of time will have a z transform in addition to a Laplace transform, and tables of z transforms can be found in many textbooks. System components are described by pulse transfer functions $G(z)$ which contain information for the sampling instants but not the time between. Since e^{-sT} represents a backward time shift T it will be appreciated that a sequence of pulses such as $e*(t)$ can be represented by a polynomial in z^{-1} whose coefficients are the pulse heights. When the input to a dynamic process is a discrete signal such as $e*(t)$ then, as a result of the filtering effect of the process, its output will be a continuous signal. Such a sequence of pulses has very little energy, so a sampler is often followed by a zero order hold device which generates a staircase waveform, as $e_h(t)$ in Fig. 4.24. Block diagrams with one or more samplers and both discrete and continuous signals at different points can be manipulated in a similar way to the approach in Section 1.5. The effect of a sampler is to tend to make a system less stable, since no account is taken of changes to the system state between sampling instants. Stability thus depends both on $G(s)$ and on T. If the sampling frequency is high, the reduction in stability is likely to be small, and vice versa. In the z plane the stability boundary is the unit circle, and if any roots of the characteristic equation lie outside this the system is unstable. As an example of the adaptation referred to above a change of variable from z to r where $z = (r+1)/(r-1)$ allows the Routh–Hurwitz technique and various frequency response methods to be used normally.

Statistical control

Analysis in this book has been limited to studying system response to, and designing for, simple deterministic forcing functions. Provided the system is linear then input signals of more complicated form can be handled by employing the principle of superposition. If the forcing function $x(t)$ is considered to be made up of a series of narrow pulses of varying height (similar to the staircase wave form in Fig. 4.24 when $T \to 0$) then the output at time t will be the sum of the response of the system to all of the pulses in the past, expressed by the **convolution integral**

$$y(t) = \int_0^t w(\lambda)x(t - \lambda)\,\mathrm{d}\lambda$$

$w(t)$ is referred to as the **weighting function** or **unit impulse response** and is the Laplace inverse of $G(s)$, since $Y(s) = G(s)$ for $X(s) = 1$.

Simple deterministic forcing functions do commonly occur in practice, although often accompanied by a noise component. If this is of small amplitude it can usually be ignored, but it may not be insignificant compared to the deterministic signal, or may be the only forcing function if the nominal input is constant. Such a signal component with random characteristics cannot be described as an explicit function of time but must be described in a statistical manner. A signal $x(t)$, whether non-deterministic or deterministic, can be described in the time domain by its **autocorrelation function** $\phi_{xx}(\tau)$ defined as

$$\phi_{xx}(\tau) = \lim_{T \to \infty} \frac{1}{2T} \int_{-T}^{T} x(t)x(t + \tau)\,\mathrm{d}t$$

It is the time average of the product of values of $x(t)$ spaced τ seconds apart, for τ varying from zero to a large value, and is a measure of the predictability of the signal. A completely random signal referred to as **white noise** has an autocorrelation function which is an impulse at $\tau = 0$, since there is no correlation between values of $x(t)$ which are τ seconds apart except when $\tau = 0$. Where there are two signals $x(t)$ and $y(t)$ then a measure of the dependence of one on the other is given by the **cross correlation function**

$$\phi_{xy}(\tau) = \lim_{T \to \infty} \frac{1}{2T} \int_{-T}^{T} x(t)y(t + \tau)\,\mathrm{d}t$$

If $x(t)$ and $y(t)$ are the input to and output from a system of transfer function $G(s)$, weighting function $w(t) = \mathscr{L}^{-1}[G(s)]$, then it can be shown that

$$y(t) = \int_{-\infty}^{\infty} w(\lambda)x(t - \lambda)\,\mathrm{d}\lambda$$

and

$$\phi_{xy}(\tau) = \int_{-\infty}^{\infty} w(\lambda)\phi_{xx}(\tau - \lambda)\,\mathrm{d}\lambda$$

A useful consequence of the similarity of these equations is that if the input function $x(t)$ is chosen to be white noise then cross correlation of the output and input functions will yield the weighting function $w(t)$. This can be an effective method for system identification, with less disturbance than step or sinusoid. A convenient approximation to white noise is a periodic signal referred to as a **pseudo random binary sequence** or **PRBS** which is a deterministic signal with random properties and has the form shown in Fig. 4.25. It has only two states, and changes in a repeatable 'random' way.

A non-deterministic signal can alternatively be described in the frequency domain by the **power spectral density** $\Phi_{xx}(\omega)$, which is the Fourier transform of the autocorrelation function, and is a measure of the energy distribution in the signal.

Fig. 4.25 Pseudo random binary sequence

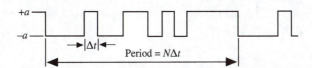

$+a$
$-a$
$\left|\Delta t\right|$
Period $= N\Delta t$

For white noise $\Phi_{xx}(\omega) = $ constant for $-\infty < \omega < \infty$, i.e. uniform energy at all frequencies. The infinite range implies infinite power, so in practice white noise is band limited and $\Phi_{xx}(\omega)$ is zero for frequencies outside the specified bandwidth of frequency.

Other control methods

An alternative to rate feedback with an auxiliary loop is **acceleration feedback**. When a process has a pair of very oscillatory poles then compensation can be effected by **pole cancellation**, the introduction of a pair of zeros at the same place, together with a pair of poles much further to the left of the imaginary axis. Cancellation cannot be exact, but the coefficient of the troublesome roots would be much reduced. **Feedforward compensation** is an approach which is helpful for systems where the point at which unwanted disturbances enter is known. If the disturbance signal can be monitored, and a mathematical model of the process is available, then the signal can be fed to a controller and then forward to the process input so that it affects the process in a way that largely cancels out the effect of the disturbance.

If a state space model is available for a process then **state vector feedback control** can be implemented, which involves feeding back the n signals in the state vector through a matrix $\mathbf{H} = [h_1 \ h_2 \ \dots \ h_n]$ to generate a process input signal $r(t) - \{h_1x_1(t) + h_2x_2(t) + \dots h_nx_n(t)\}$. There are n parameters whose values should be chosen to achieve suitable dynamic behaviour (compared to 1, 2 or 3 for P, PI/PD or PID control). Limitations are that all of the n state variables may not be available, a problem of observability, and that certain variables may not be controllable. **Adaptive control** is an approach which can help with processes whose dynamic characteristics are slowly time varying. It involves measurement of the dynamics followed by adjustment of the controller parameters so that some chosen performance criterion is continuously optimised. It could take the form of automatic tuning of a PID controller.

Problems

Problems 1 to 7 all refer to closed loop systems where $G(s)$ is the forward loop transfer function and $H(s)$, if not unity, is a transfer function in the feedback loop, as in Fig. P4.1(a). In problems 8 to 12 and 15 to 17 there is, in addition, a controller or compensation network with transfer function $G_c(s)$ in series with the process $G_p(s)$ being controlled (Fig. P4.1(b)).

1   Sketch, making use of the appropriate aids to construction, the general form of the root locus plots which have the open loop transfer functions $G(s)H(s)$ given below, indicating the key features. Outline what can be deduced about the system behaviour.

Fig. p4.1

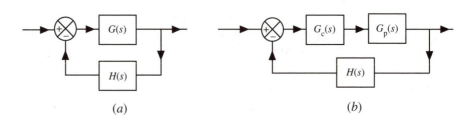

                           (a)   (b)

(a) $G(s)H(s) = K/(s+2)(s+10)$
(b) $G(s)H(s) = K/s(s^2+5s+25)$
(c) $G(s)H(s) = K(s+2)/(s+1)(s+6)$
(d) $G(s)H(s) = K/s^2(1+5s)$
(e) $G(s)H(s) = K(1+2s)/s(1+0.5s)(s^2+s+9)$

2   A unity feedback system has forward loop transfer function $G(s) = K(s+1)/(s^2+4s+9)$. Using the aids to construction sketch the general form of the root locus plot. By drawing to scale and using the angle condition to determine selected intermediate points draw the plot more exactly. Use the magnitude condition to identify values of K along the plot. Check selected points by solving the characteristic equation.

3   A unity feedback closed loop system has forward loop transfer function $G(s) = K/(s+1)(s^2+4s+5)$. Sketch the form of the root locus plot. What is the limiting value of K for stable operation? Does it appear possible to achieve a value of 0.7 for the damping factor of the dominant roots? If so draw the appropriate part of the plot more accurately and determine the value to which the gain K should be set.

4   A closed loop system has open loop transfer function $G(s)H(s) = K(s+2)/s(s-1)(s+8)$. Sketch the root locus plot and explain what the plot indicates about the nature of the dynamic behaviour of the loop.

5   For a unity feedback closed loop system with $G(s) = K(s+\alpha)/s^2(s+10)$ investigate how variation in the value of α alters the form of the root locus plot by considering in turn $\alpha = 0, 0.5, 1, 2, 5$ and 10.

6   For the system of Problem 5 draw root locus plots to indicate the variation of the characteristic equation root positions in the s plane with variation of the parameter α for the two cases of $K = 10$ and $K = 50$.

7   A process, whose output is controlled by means of a feedback loop, has transfer function $G(s) = K_1/s(1+0.1s)$ and the transducer in the feedback loop has transfer function $H(s) = K_2/(1+Ts)$. Draw a root contour plot showing the nature of the variation of the roots of the characteristic equation with varying T for values of $K_1K_2 = 20, 5, 2$ and 1.

8   A process with transfer function $G_p(s) = 1/(1+s)(1+0.5s)$ is to be controlled by means of a proportional plus integral controller with transfer function $G_c(s) = K(1+k_i/s)$ in the forward loop of a closed loop system with unity feedback, $H(s) = 1$ (K includes the process gain). On a scale drawing sketch a set of root contours showing how the roots of the characteristic equation alter with k_i for selected values of gain K, and hence suggest values of K and k_i which could be expected to yield a response with a value of 0.7 for the damping factor of the dominant roots.

9   Repeat Problem 8 for a process with transfer function $G_p(s) = 1/(s^2+5s+9)$.

10 A P + I controller with transfer function $G_c(s) = K_c(1+k_i/s)$ is used for speed control of a motor with transfer function $G_p(s) = K_m/(1+T_m s)$. The transducer in the feedback loop is a pure gain, $H(s) = K_t$. Construct a root contour plot to show the variation in the root positions for changing K_c and k_i, when the parameters are $K_m = 20$, $K_t = 0.01$, and $T_m = 0.5$s. What can be deduced from the plot?

11 Draw a root contour plot for Example 4.6 to show the variation of root position with k_d for a range of values of K.

12 A process with transfer function $G_p(s) = 1/s(1+s)(1+5s)$ is to be controlled by a proportional plus derivative controller with transfer function $G_c(s) = K(1+k_d s)$ preceding the process in the forward path of a unity feedback closed loop system. Sketch a root contour plot to show the effect on the characteristic equation roots of variation of k_d for different values of K. Explain what can be deduced about the dynamic behaviour which can be expected.

13 How would the root contour plot for the negative velocity feedback system of Example 4.7 differ if the motor time constant were 0.1 second instead of 1 second? How can the response be expected to differ?

14   The use of a unity feedback loop together with a secondary rate feedback loop (Fig. 4.15) is to be investigated for control of the output of a process with transfer function $G(s) = K/s(s^2 + 5s + 9)$. Draw a root contour plot to show the variation of the system roots with change in K and k. Describe the nature of the transient behaviour which can be expected for different values of these two parameters and suggest a combination of values which might be appropriate to give a relatively fast and well damped response.

15   Design a phase lead compensation network for a unity feedback system with forward loop transfer function $G(s) = K/(1 + 0.2s)(1 + 0.5s)$ in order to achieve a positional error of not greater than 2% and a phase margin of at least 50°.

16   Design phase lead compensation for the control of a process with transfer function $K/s(1 + 0.01s)(1 + 0.03s)$ in a unity feedback loop. The performance requirement is that the velocity error coefficient K_v should have a value of 40 and the phase margin should be at least 50°.

17   A process has transfer function $K/s(1 + 0.1s)(1 + 0.5s)$ and is to be controlled by means of a unity feedback loop and a series phase lag compensation network. Determine an appropriate transfer function for the compensation network to satisfy a specification of (a) $K_v = 20$, phase margin = 45°; (b) $K_v = 10$, phase margin = 60°. Describe qualitatively how you would expect the transient response to differ in the two cases.

BIBLIOGRAPHY

Listed below is a selection of textbooks which the author feels would serve the reader well as complementary texts which provide more detail, different perspectives, and coverage of more advanced topics.

Schwarzenbach J and Gill KF *System Modelling and Control*, Edward Arnold, 3rd Edition, 1992. 337 pages. Provides somewhat more detail in certain areas and rather less in others, as compared with the present book. It includes chapters on System Simulation, State Space Representation and Analysis, The Sampled Data Process, and Nonlinearities and Dead Time.

Golten J and Verwer A *Control System Design and Simulation*, McGraw Hill, 1991. 384 pages. Broadly similar coverage in content and level of difficulty. Also includes chapters on Discrete Time Systems, Computer Control Systems, and Nonlinearities in Control Systems. Available with the book on disk is a very useful interactive graphics software package (CODAS) for design and simulation of control systems. This is the source of the graphical output for the worked examples in the text.

Richards RJ *Solving Problems in Control*, Longman, 1993. 230 pages. Covers the same topics as this book by summarising the essential theory and then demonstrating the application of the theory by worked examples which solve examination type questions.

Kuo BC *Automatic Control Systems*, Prentice Hall, 7th Edition, 1995. 865 pages. A well established comprehensive text with in-depth coverage of linear continuous data and digital control systems analysis and design. A supporting control systems software package is available on disk, together with a manual.

Nise NS *Control Systems Engineering*, Benjamin/Cummings, 2nd Edition, 1995. 837 pages. Very comprehensive coverage primarily treating the topics in this book, but also including State Space and Digital Control Systems. Each chapter incorporates a case study, and the use of MATLAB for computer-aided analysis and design is integrated into discussions and problems as an optional feature.

Dorf RC and Bishop RH *Modern Control Systems* Addison-Wesley, 7th Edition, 1995. 800 pages. A comprehensive text which in addition to the topics covered in this text has chapters on State Variable Models, The Design of State Variable Feedback Systems, Robust Control Systems, and Digital Control Systems. Particular features are the large number of practical illustrations in the text and in the problems, the inclusion of MATLAB programs for obtaining the different forms of plot, and problems in each chapter grouped into 5 different categories.

ANSWERS TO PROBLEMS

Chapter 1

1 Figure A1

Fig. A1

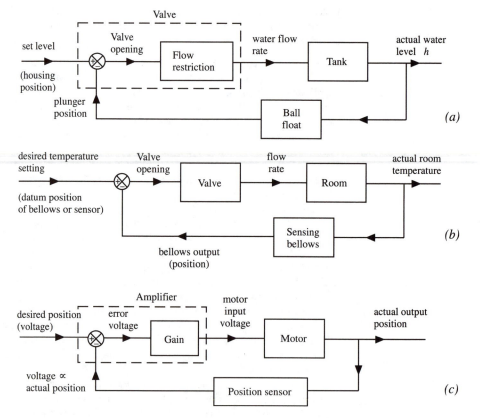

2 $\dfrac{5}{s^2}$; $\dfrac{10}{s}+\dfrac{2}{s^2}$; $\dfrac{1}{s}+\dfrac{1}{s+3}$; $\dfrac{1}{s}(1-e^{-10s})$

3 $(s^3+5s+1)C(s)$; $\dfrac{s+2}{s^2+4s+4.25}$; $\dfrac{5}{s}+\dfrac{20}{s^2}$; $X(s)=10sY(s)+5Y(s)$; $C(s)=\dfrac{2\omega}{s^2+\omega^2}$

4 $3e^{-4t}$; $\frac{2}{3}(1-e^{-3t})$; $2.5(e^{-2t}-e^{-10t})$; $0.2-0.25e^{-3t}\sin(4t+53.1°)$;
$0.2t-0.048+0.165e^{-3t}\sin(3t+16.9°)$

5 0; $2/3$; 0; 0.2; ∞

6 $1/(ms^2+cs+2k)$

7 $1/(ms^2+cs+2k_1)$; $k_2/[ms^2+cs+(2k_1+k_2)]$

8 $(J_2s^2+C_2s+K)$ /s $[J_1J_2s^3+(J_1C_2+J_2C_1)s^2+(J_1K+J_2K+C_1C_2)s+(C_1K+C_2K)]$;
$K/s[J_1J_2s^3+(J_1C_2+J_2C_1)s^2+(J_1K+J_2K+C_1C_2)s+(C_1K+C_2K)]$;
$1/s[(J_1+J_2)s+(C_1+C_2)]$

9 $1/s[(C_1+C_2/n^2)+(J_1+J_2/n^2)s]$

10 $1,(R_1+R_2)C,R_2/(R_1+R_2)$; $1/\alpha,R_1R_2C/(R_1+R_2),(R_1+R_2)/R_2$

11 $(1+R_1C_1;s)(1+R_2C_2s)/[1+(R_1C_1+R_2C_2+R_1C_2)s+R_1R_2C_1C_2s^2]$

12 $1/[1+(A_1R_1+A_2R_2+A_1R_2)s+A_1A_2R_1R_2s^2]$;
$R_2/[1+(A_1R_1+A_2R_2+A_1R_2)s+A_1A_2R_1R_1s^2]$;
$(R_1+R_2+A_2;R_1R_2s)/[1+(A_1R_1+A_2R_2+A_1;R_2)s+A_1A_2R_1R_2s^2]$

13 $G_1G_2G_3G_4/[1+G_1G_2G_3G_4+G_3G_4H_1+G_2G_3G_4H_2]$

14 $G_1G_3G_4(G_2+G_5)/[1+G_3G_4(G_2+G_5)+G_1G_3H(G_2+G_5)]$

15 $1/K_p,(K_mK_aK_t+1)/2\sqrt{K_mK_aK_pT_m}$; $\sqrt{K_mK_aK_p/T_m}$;
0.625 r/volt, 0.59, 3.58 r/s; $\ddot\theta_m(t)+4.22\dot\theta_m(t)+12.82\theta_m(t)=8.01v_i(t)$

Chapter 2

1 $1-e^{-10t}$; $10e^{-10t}$; 0.39 sec (to within 2%)

2 $t-2+2e^{-0.5t}$; $5t-10+10e^{-0.5t}$; 2, 10; $\times 4$

3 $0.5-0.75e^{-2t}+0.25e^{-6t}$; $5-7.5e^{-2t}+2.5e^{-6t}$; $1/2t-1/3+3/8e^{-2t}-1/24e^{-6t}$; 0, 0, $1/3$
(gain is 0.5)

4 $1-1.091e^{-2t}\sin(4.58t+1.16)$; 25.4%

5 $120-13.3e^{-2t}+60e^{-t}-166.7e^{-0.2t}$ (both)

6 $2.5-2.5e^{-t}-1.29e^{-0.5t}\sin 1.936t$; $2.5-2.558e^{-0.1t}-0.130e^{-0.5t}\sin(1.936t-0.460)$

7 $0.500\underline{/-5.7°}$, $0.496\underline{/-14.4°}$, $0.486\underline{/-29.0°}$, $0.429\underline{/-59.0°}$, $0.171\underline{/-121.0°}$,
$0.049\underline{/-151.0°}$, $0.012\underline{/-165.6°}$

8 $\omega=0$ $1\underline{/0°}$; $\omega=0.2$ $0.700\underline{/-56°}$; $\omega=0.5$ $0.347\underline{/-95°}$; $\omega=1$ $0.156\underline{/-129°}$;
$\omega=2$ $0.052\underline{/-168°}$; $\omega=4$ $0.011\underline{/-207°}$; $\omega=\infty$ $0\underline{/-270°}$

9 $\omega=0.1$ $6.9\underline{/-147°}$; $\omega=0.2$ $2.1\underline{/-176°}$; $\omega=0.3$ $0.9\underline{/-194°}$; $\omega=0.5$ $0.28\underline{/-217°}$
$1/s(1+2s)(1+10s)$ gives very similar response, except at high frequencies

10 As problem 9

11 Figure A2

12 Figure A3

Fig. A2

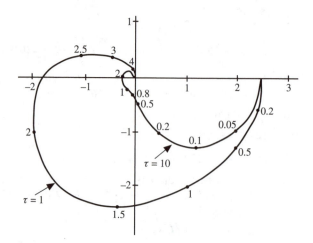

Fig. A3

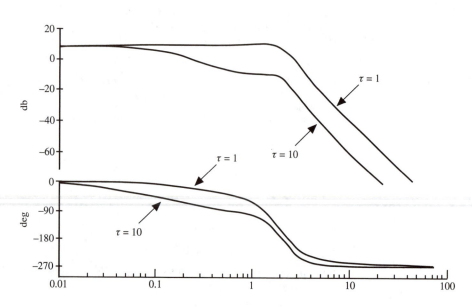

13 $1 - 1.054e^{-t} \sin (3t + 1.25)$

ω(r/s)	0	1	2	3	4	5	6	8	10
Mag	1	1.084	1.387	1.644	1.000	0.555	0.349	0.178	0.109
Phase (deg)	0	− 13	− 34	− 81	− 127	− 146	− 155	− 164	− 168

14   Roots at $- 0.2 \pm j0.98$ dominant, root at $- 5$ insignificant. Response close to second order with $\omega_n = 1$, $\zeta = 0.2$, gain 2, i.e. peak overshoot $\approx 2 \times 1.55$, period of oscillation just less than 2π sec, settling time $\approx 4/0.2$ sec, settling to 2. Effect of lag(s) is to reduce overshoot a little, and delay whole trace a little

15 $2 - 2e^{-0.2t}$ for $0 \le t \le 4$; $6 - 10.9e^{-0.2t}$ for $t \ge 4$

16 $t - 6 - 0.25e^{-t} + 6.25e^{-0.2t}$ for $0 \le t \le 10$
(a) $10 + 5506e^{-t} - 39.93e^{-0.2t}$ for $t \ge 10$
(b) $3t - 38 + 98.6e^{-0.2t} - 1103e^{-t}$ for $t \ge 10$

17 $1/4 - 1/3e^{-t} + 1/12e^{-4t}$ for $0 \leq t \leq 5$

$(1/4 - 1/3e^{-t} + 1/12e^{-4t}) - (1/4 - 1/3e^{-(t-5)} + 1/12e^{-4(t-5)})$ for $5 \leq t \leq 10$

$(1/4 - 1/3e^{-t} + 1/12e^{-4t}) - (1/4 - 1/3e^{-(t-5)} + 1/12e^{-4(t-5)})$

$+ (1/4 - 1/3e^{-(t-10)} + 1/12e^{-4(t-10)})$ for $10 \leq t \leq 15$

Figure A4

Fig. A4

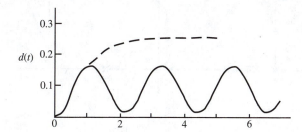

Chapter 3

1 (a) -0.5, -1.2, -2; overdamped, dominant $\tau = 2$; settling $> 4 \times 2$, say 10 sec

(b) -2, $-1 \pm j1.414$; latter dominant, $\zeta \approx 0.57$, ≈ 2.3 sec to peak; probably just 1 overshoot

(c) -3, $1 \pm j2$; unstable; slight disturbance: oscillation ≈ 2 r/s increasing in amplitude

(d) -1, -3, $-1 \pm j1.73$; -1 and complex roots equally dominant; overshoot much less than 15%, settling time somewhat greater than 4 seconds

2 a, b, d stable; c unstable

3 Yes, 2

4 Stable for $K < 4.65$; 0.2 r/s

5 Stable for $K < 390$; 2.37 r/s

6 $215 < K < 2378$; 2.72 r/s, 9.91 r/s

7 Stable; 5.8 db, 25°

8/9 2.4 db, 20°; reduce gain by 1.60, 6.5 db; reduce gain by 1.52, 48°

10 0.4 db, 3°; reduce gain by 2.2, 7.0 db; reduce gain by 1.9, 45°

11/12 $M_p \approx 4$, $\omega_p \approx 2.75$ r/s, bandwidth ≈ 3.7 r/s

13 $M_p \approx 8.5$ db ≈ 2.7; $\omega_p \approx 3.1$ r/s; bandwidth ≈ 4.1 r/s; reduce numerator to 25, 3.5 r/s

14 0; 0.64; 0.13

15 $1/(1+K)$ (unit step), ∞ (ramp); no change

16 125; 0 (step), 1/5 (unit ramp), ∞ (accel)

Chapter 4

1   Figure A5

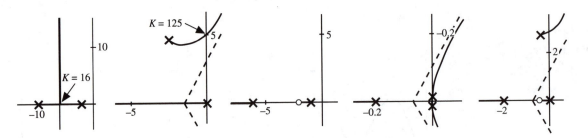

Fig. A5

2   Figure A6

Fig. A6

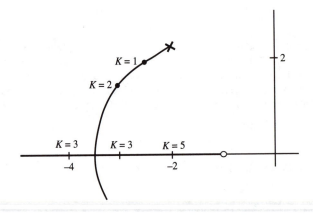

3 $K_{\text{crit}} = 40$; $\zeta = 0.7$ for $K \approx 2.46$

4   Figure A7. Unstable at low K. Stable for $K > 11, 2$. Max $\zeta \approx 0.35$

Fig. A7

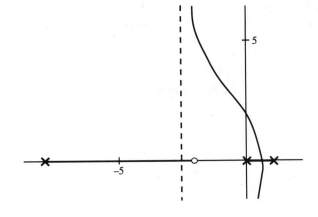

5   Figure A8

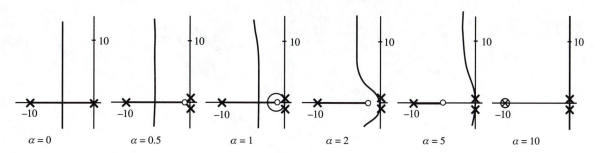

Fig. A8

6   Figure A9

Fig. A9

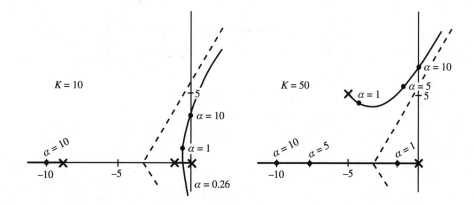

7   Figure A10

Fig. A10

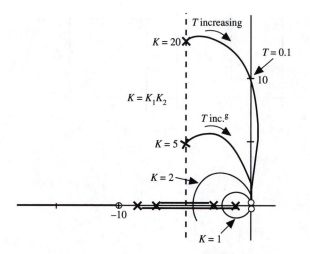

8   Figure A11. $K = 1$, $k_i = 1.05$

Fig. A11

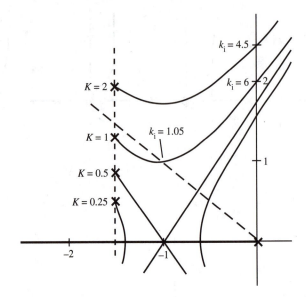

9 $K = 2$, $k_i = 4.7$ or $K = 1$, $k_i = 8.6$

10   Figure A12

Fig. A12

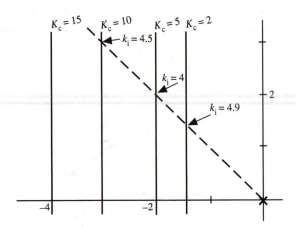

11 Figure A13

Fig. A13

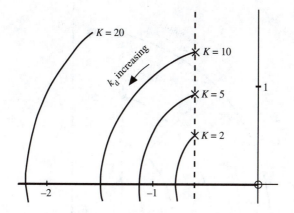

12 Figure A14

Fig. A14

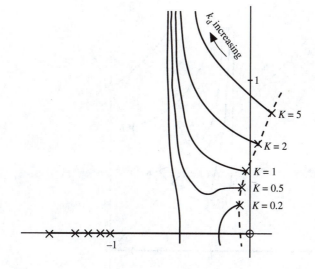

13   Figure A15. Faster by factor ≈ 5; $K \approx 2\times$, $k \approx 0.1\times$

Fig. A15

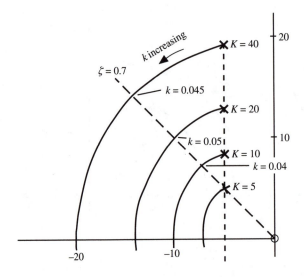

14   Figure A16. $K = 10$, $k = 0.25$ for $\zeta \approx 0.7$ for dominant roots

Fig. A16

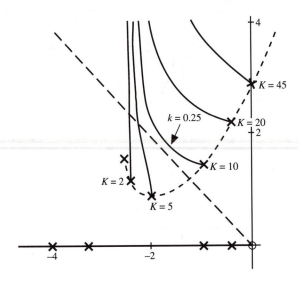

15 $K = 200$, $G_c(s) = \dfrac{1 + 0.064s}{4(1 + 0.016s)}$

16 $K = 146.8$, $G_c(s) = \dfrac{1}{3.67} \dfrac{1 + 0.044s}{1 + 0.012s}$

17   (a) $K = 20$, $G_c(s) = \dfrac{1 + 7.7s}{1 + 100s}$; (b) $K = 10$, $G_c(s) = \dfrac{1 + 13.4s}{1 + 167s}$

Index